Fractal Geometries
Theory and Applications

Alain Le Méhauté
Translated by Dr Jack Howlett

(CRC)

CRC Press Inc.
Boca Raton - Ann Arbor - London

© Hermes, Paris 1990
English translation © Penton Press, London 1991

First published in French as *Les Geometries Fractales: L'espace-temps brisé* in 1990 by Editions Hermes, 34, rue Eugène Flachat, 75017 Paris.

First published in English in Great Britain by
Penton Press Ltd, London.

First published in English in the United States by
CRC Press Inc
2000 Corporate Blvd N.W.
Boca Raton
FL 33431, USA

ISBN 0 8493 7722 6

Library of Congress Cataloging in Publication Data
Géometries fractales. English
　Fractal geometries / edited by Alain Le Méhauté: translated by Jack Howlett.
　p.　cm.
　Translation of Les Géometries fractales.
　Includes bibliographical references and index.
　ISBN 0 8493 7722 6
　1. Fractals.　I. Le Méhauté, Alain.　II. Title.
QA614.86.G46　1991
514　.74--dc20

BK
$30.00

Printed and bound in Great Britain

Contents

Foreword

It was in 1973 that the author chanced upon the work of Benoit Mandelbrot, author of *FRACTALS* and subsequent books: these, mapping out the previously uncharted area of fractal geometry, came as a revelation. Perhaps surprisingly, among the first to realise the practical importance of this were scientists working as research engineers in industry, for example in Exxon, Union Carbide, Brown Bovery and the Compagnie Générale d'Electricité, all interested in practical applications.

In elementary mathematics we learn about sets bounded by smooth curves, which are all that seem to be needed. The mathematician describes these boundaries as *rectifiable*, meaning that they are locally straight. Going from here to frontiers and coasts we conclude that these are of finite length, and that singularities such as islands and peninsulas can be counted by ordinary finite numbers. All phenomena that follow are differentiable and vector operators can be applied.

This simple view has been modified by the work of Mandelbrot, who realised that natural sets, although bounded, are not finite. He saw the importance of this and understood its geometry, to which he gave the name *fractal*; he and his followers raised the science to the level of a language, based on the generic property of invariance under iteration. But there can be no iteration without the passage of time: time results from geometrical iteration and runs irreversibly most often when space has lost its boundary.

Mandelbrot was interested in frontiers, and indeed the present book also is concerned with frontiers. He has classified as *fractals* these fuzzy and tenuous limits of unbounded space, the principal characteristic of which is a *fractal dimension*. This abstract measure, due to Hausdorff, results from a construction by iteration; the concept goes back at least as far as Leibnitz.

The first chapter of the book sets out the elementary procedure that leads to the concept of non-integral dimension. This will enable the reader to understand, without the need for any mathematical tools, why coastlines and frontiers, regions of exchange *par excellence*, can be fuzzy; and to create such frontiers. The simplicity of the properties whose derivation is the main object of the analysis should stimulate his imagination and at the same time lead him to question much that he learned at school or university. Deliberately ignoring topological considerations, the descriptive parameters of the new geometry are reduced to two: the dimension, a state parameter, and the co-dimension, a parameter of the environment. This is indeed a simplification, and therefore one has to go further, to the start of Chapter 2, to understand that there are several ways of arriving at the value of the dimension parameter, and therefore several dimensions. Some details are necessary here, which are quickly dealt with. The second part of this chapter builds on the previous arguments; here we find, using the results of Charles Tricot, another very talented mathematician, that the parametrisation of a curve underlies the space-time coupling of physics and in fact controls the irreversibility features which put their stamp on the world of our experience. Irreversible time is a consequence of the exigencies of parametrisation and the constraints of uncertainty. The concept of velocity is generalised.

The different modes of parametrisation constrain the action but create possibilities of unlimited extent. We shall show that ergodic properties in a fractal environment can be based on either the dimension of the space or on its co-dimension. This leads to a duality between dimension and co-dimension analogous to that between dynamical variables and the conjugated variables. Two modes of irreversibility are then distinguished, according as the fractal determines the limits on physical processes embedded in a Euclidean space, or the correlations assign the contents 'themselves of dissipative spaces. The analysis leads naturally to the question of what is the operator that is symbolic of such irreversibilities? The answer has to be found in the theory of distributions, in particular in the smoothing that results from convolution procedures under the effect of the measure. Here the relevance of the operation of fractional derivation, the simplest operator of space-time coupling, becomes evident. Chapter 3 establishes the main properties of this operator and relates them to the fractal character of the space, in particular to its non-integral dimension. It is shown here in passing that the fractional derivation operator optimises the exchanges and minimises the dissipations.

With multifractality, conceived by Mandelbrot between 1968 and 1976, fractality becomes enriched. The architecture becomes baroque, multiple and complex, more narrowly conditioned to meet the demands of the real world. Chapter 4 outlines a multifractality which is fully developed in Mandelbrot's latest publications. After the unit of space has been defined by means of the multifractality, it becomes necessary, for practical reasons, to define the unit of time. How can we explain the irreversibility that we perceive in a space of the complexity resulting from the mixture of scales and dimensions? Our analysis takes the approach of multifractality based on probabilities. The physical applications we describe demand a definition of *hyperfractality*: this, like multifractality, is a matter of overlaying fractal sets, but the source of the concept is in the passing of time and not in the measurement of space. Three elementary examples are used to illustrate the work of this chapter.

Descriptive analysis can seem abstract, and it is true that the network of relations flowing from fractality is still not fully explored. However, Part II, *Applications*, gives some applications of the language that has been developed and the images reproduced here will help to illuminate the field of the analysis. Most of these are original and I record my grateful thanks to those who generated them. In a way, they reveal a new form of the fantastic, the sharing of which, over and above the scientific interest, has been my main motivation in writing this book.

Action is the reason for writing this book. It is the work of an engineer who is convinced that nothing can be done without understanding, by which I mean that understanding must precede creative action.

Notation

ω	:	frequency.
$p\,(= i\omega)$	:	Laplace generalised frequency, related to time t.
Δ	:	fractal (self-similarity) dimension.
Δ_{MB}	:	Minkowski-Bouligand dimension.
ζ	:	co-dimension.
$\xi(t)$	:	gauge in space-time measure, associated with the parametrisation γ.
$l(t)$	:	length associated with gauge $\xi(t)$.
$\eta(p)$	:	gauge in Laplace or Fourier space.
$\lambda(p)$	:	length associated with gauge $\eta(p)$.
$\varphi(t)$	:	test function, flux density.
$J(t)$	:	flux.
$\Delta X(t)$	:	force increment.
$C(p)$	:	state function associated with a fractal interface.
p_r	:	probability.
D_{iff}	:	normal diffusion coefficient.
D_{iffH}	:	abnormal diffusion coefficient.
$a\mathscr{D}_t^{\alpha}$	:	fractional derivation operator of order α with respect to t. a is the lower limit of the convolution integral.
$*$	:	convolution: $f(t) * g(t) = \int f(u)g(t-u)\,du$.
$\sim$	:	"behaves like", often used in a very general sense: $f(t) \sim g(t)$ means that f and g have the same general behaviour.

Part I

Theory and Techniques

Chapter 1

The discovery of fractal geometry

The aim of this chapter is to give the reader a short summary of the fundamentals of fractal geometry. There are, of course, excellent works on this, some of which are listed in the bibliography, which will take the reader further with the analysis that is only sketched out here. All we attempt here is an elementary presentation of the relations that have to be known in order to follow, without difficulty and with a critical mind, the scientific literature dealing with *the concept of time in physics in a fractal environment*. We have preferred clarity to mathematical rigour. In this study the reader will find that the key element of fractal geometry, the non-integral dimension, is intimately associated with the way in which we evaluate our space, measure its boundaries and "weigh" its contents; the meaning of the concept of *dimension* will be seen to be, above all, physical. He will see also that the aim of a *measure* is to try to free the mind of the paradoxes that characterise "pathological" behaviour, whilst using examples of such behaviour to enrich our vision of the world. Finally, he will see that this freedom can be exercised at the mathematical level, perhaps more there than elsewhere – something that the engineer too often forgets.

1.1 The surveyor's task: rectifiable curves, measuring by arc lengths

For most human beings the length of a curve is a primitive notion, acquired in the first years of life. Doubtless Neanderthal man, although never having formalised the knowledge, knew that the "longer" a path the more time it would take him to follow it; and similarly he would know that it was easier to walk on the flat than up a mountain. Even so, he would not know the meaning of *rectification*, a term invented much later to express our ability to measure the

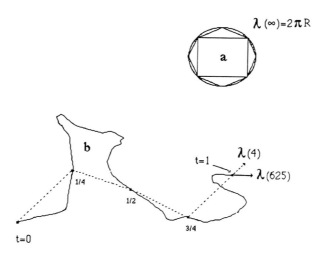

Figure 1.1 Approximation to a curve by steps of length η, for $p = 4$ and $p = 625$. We know that the perimeter of the circle is the finite number $\lambda = 2\pi R$, where R is its radius; this is the value obtained when $\eta \rightarrow 0$.

lengths of smooth curves. Consider this concept for a moment: let us call a certain path an arc Γ, and suppose that a person walks along this at a "speed" expressed by a certain function v(t). Each point on Γ will be determined by a function l(t) which gives the distance travelled from the starting point, where "t" is the "time" of arrival at that point. We say that the curve is *parametrised* by l(t), which initially we assume to be defined and continuous everywhere. This will be the case if the walker does not jump erratically from place to place, and indeed the principle of the walk requires that he does not. The *step* is the human gauge for measuring lengths along the paths we take. But even without any jumping about the walker will not follow the curve in all its infinitesimal details; implicitly, he will approximate it by a polygonal line whose general form is given by steps of length η placed on the ground in succession.

Let $\lambda(\mathscr{P})$ be the length of the plane polygonal line $[f(0), f(t_1), f(t_2), \ldots f(t_{p-1})]$, where $f(t_i) = [x(t_i), y(t_i)]$ and $\mathscr{P}$ is the finite sequence $[t_1, t_2, \ldots t_p]$. We say that the arc is rectifiable if the upper bound of the real numbers $\lambda(\mathscr{P})$ when $p \rightarrow \infty$ is a finite number L, which is then called the length of the curve Γ.

So far no constraint has been placed on the elementary segments that constitute the polygonal line; suppose now that these are all of the same length $\eta(p)$ and that all the time intervals $t_i - t_{i-1}$ are equal and have the value Δt. Then the sequence $\mathscr{P}$ is equivalent to giving the number p of measuring steps. If t is the total time needed to make the measure, $p = t/\Delta t$; and if $p = p/\tau_0$ where τ_0 is some constant time then p can be regarded as a generalised frequency and used as a Laplace variable. We have now

$$\lambda(p) = N(p) \cdot \eta(p) \tag{1.1}$$

where N is the number of steps in the polygonal line. Rectifiability can be understood simply from a graph of $\lambda(p)$ as a function of $\eta(p)$: it means that $\lambda(p)$ must tend to a finite limit, L, as $\eta(p)$ tends to zero. The necessary analysis can be performed in Laplace or Fourier space.

$\lambda(p) \rightarrow L$ when $p \rightarrow \infty$, that is when $\Delta t \rightarrow 0$ and t is finite

Stated otherwise, the space gauge tends to zero when the time gauge tends to zero, or equivalently when the frequency gauge tends to infinity: an infinitely precise measure requires an infinite frequency.

Note that if p is the generalised frequency defined above, the number of steps per unit time is proportional to p and also $N(p) \sim p$.

The concept of rectifiability is simple, intuitive and almost natural, but does it hold universally? Ten years ago almost everyone except for a few mathematical specialists would have said yes. However, as Mandelbrot recalled in 1975, the true answer is no; a study of Figure 1.2 should put us on our guard. Approximating a rectifiable curve assumes some very special properties. There

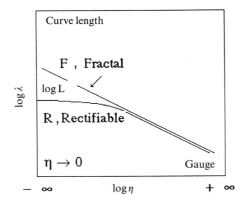

Figure 1.2 Characteristics of rectifiable (R) and fractal (F) curves. For R the length $\lambda(\eta)$ measured in steps of length η approaches a finite limit L as $\eta \rightarrow 0$, for F there is no limit. The concept of non-integral dimension is related to the slope of a fractal curve.

are curves with strange properties, such that the length does not tend to a finite limit as the step length $\eta(p)$ tends to zero. How can we classify these? What properties are hidden in them? The essential aim of this chapter is to show that the concept of "measure", in the physical sense of the term, enables us to bring the properties of these strange objects under our control. But we shall now leave further discussion of the parameter p to the next chapter, and concentrate on the analysis of the space.

During the last century mathematicians established the formal existence of certain pathological curves: for example, Cantor, Peano, Hausdorff, Bouligand among others. But it was the expatriate French geometer Mandelbrot who first understood the depth of this work and made it more widely known, in particular characterising the strange objects by the term *fractal*. This was not easily achieved; his first book, published in 1975, was effectively ignored and he had to wait until 1983 for recognition of the importance of his contribution. Eight years after the birth of his ideas, and thanks to a generation of young physicists led by Mandelbrot himself, the importance of these pathological objects has at last been acknowledged by the great majority of the scientific community.

1.2 A journey into pathology: the search for lost rectifiability

Let us look more closely at Figure 1.2. There is nothing implied here concerning the function that expresses rectifiability: it says that R approaches a finite limit as $\eta(p)$ tends to zero and F does not, but says nothing about the form of F. Let us now *assume that F is straight line with slope $1 - \Delta$*: this is a somewhat draconian assumption, but we take it for what it is, a provisional heuristic – which we shall see later to be an excellent opening move.

The linearity of F in the $[\log \eta(p), \log \lambda(p)]^*$ plane implies

$$\lambda(p) \sim [\eta(p)]^{1-\Delta} \tag{1.2}$$

Since $\lambda(p)$ is just the length of the polygonal line used for rectification of the curve, $\lambda(p) = N(p) \cdot \eta(p)$ by 1.1 and so

$$N(p) \sim [\eta(p)]^{-\Delta}$$

---

* Throughout this book "log" will mean logarithm to base e

This is the formula given by Mandelbrot in 1975:

$$N(p) \cdot [\eta(p)]^{\Delta} = l_0{}^{\Delta} \tag{1.3}$$

where l_0 is a characteristic length and $N(p)$ is the number of steps of length $\eta(p)$ in the approximation to the curve.

The terms $[\eta(p)]^{\Delta}$, $l_0{}^{\Delta}$, of dimension $[L]^{\Delta}$, thus relate to spatial properties which are strange compared to those of length, area and volume with which we are familiar, since the parameter Δ, determined by the slope of the line F, is not necessarily an integer as it is in ordinary space. If $\Delta = 1$, equation 1.3 expresses the measure of a line of length l_0, which is not surprising since stating $\Delta = 1$ is equivalent to stating that the curve Γ is rectifiable. However, if $\Delta = 2$ the equation expresses the measure of an area tiled with small squares of side $\eta(p)$: this is much more surprising, because it seems that we have arrived at the measure of a surface by means of a procedure that uses the length of an arc. This observation, already made by Peano in 1890, also contains some of the strangeness of equation 1.3.

We now check that 1.3 expresses the metrical properties of ordinary space; that is, that Δ is a spatial dimension in the usual sense of the term.

(i) If $\Delta = 1$ and $l_0 = 1$, then 1.3 gives the measure of a line of unit length in terms of a gauge of length $\eta(p)$, this measure being the number of steps $N(p)$. If the gauge length is $1/2$ the measure is $N(p) = 2$; if it is $1/4$, $N(p) = 4$, and so on; and 1.3 states

$$2.(1/2)^1 = 1, \quad 4.(1/4)^1 = 1, \text{ etc.}$$

(ii) If $\Delta = 2$ and $l_0 = 1$ then $N(p) \cdot [\eta(p)]^2 = 1$ and a gauge length $\eta(p)$ implies the use of a gauge of area $[\eta(p)]^2$ for measuring the content. It follows from the equation that if $\eta(p) = 1/2$, $N(p) = 4$; if $\eta(p) = 1/3$, $N(p) = 9$, and so on:

$$9.(1/3)^2 = 1, \quad 81.(1/9)^2 = 1, \text{ etc.}$$

(iii) If $\Delta(p) = 3$ and $l_0 = 1$, $N(p) \cdot [\eta(p)]^3 = 1$ and a gauge length $\eta(p)$ implies the use of a gauge of volume $[\eta(p)]^3$ for measuring content. In this case, if $\eta(p) = 1/2$, $N(p) = 8$; if $\eta(p) = 1/3$, $N(p) = 27$ and so on:

$$8.(1/2)^3 = 1, \quad 27.(1/3)^3 = 1, \text{ etc.}$$

We express these results by saying that the line is of dimension 1, surface of dimension 2 and volume of dimension 3. However, whilst the interpretation of

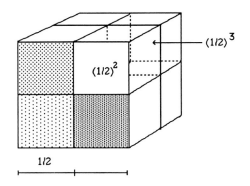

Figure 1.3 Classical structure of objects of dimension 1, 2, 3.

equation 1.3 is natural and easily understood when the dimension Δ is an integer, the derivation implies no such restriction and Δ can be given any value. This is equivalent to saying that the linearity of F defines the metric of a space whose dimension can be any real number, and leads us to consider the properties of such spaces.

To take a specific example, let us try to construct a mathematical object such that equation 1.3 applies with

$$\Delta = \log(N)/\log(1/\eta) = \log 4/\log 3$$

The expression for Δ means that if $l_0 = 1$ then for all values of p the equation is satisfied by $N(p) = 4$, $\eta(p) = 1/3$; and this means that the gauge used to rectify the curve is such that for an arc of any length, four steps of length equal to one-third of the arc length are needed. The curve to which this corresponds, the so-called von Koch curve, is shown in Figure 1.4. It is generated by the broken line of Figure 1.4a and for the successive approximations we have:

(1) $4.(1/3)^\Delta = 1$ and $N(3) \cdot [\eta(3)]^\Delta = 1$

(2) $16.(1/9)^\Delta = 1$ and $N(3^2) \cdot [\eta(3^2)]^\Delta] = 1$ etc.

The successive stages are characterised by having an increasing number of singular points and a certain similarity one with another. No limit is implied in the procedure and the iteration can be continued indefinitely. As $\eta(p) \to 0$ the structure created converges towards a curve Γ_F which is of the type called *fractal* by Mandelbrot, from the Latin *fractus*, broken; it consists of a

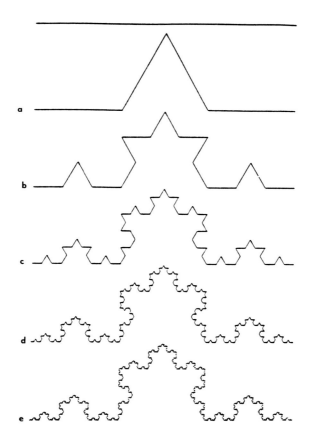

Figure 1.4 Construction of the von Koch curve by repeating the generator 1.4a on each base element and continuing the iteration indefinitely.

continuous set of self-similar singularities such that any part of the limit set is similar to the whole. This is one of the main properties of such a set.

The mathematician von Koch devised this curve in 1903 as an example of a curve that was everywhere continuous but nowhere differentiable. It was immediately relegated to the class of pathological concepts, of no interest to anyone but a few mathematicians. It is to Mandelbrot's great credit that in 1975 he brought such curves to light and used them to spread awareness of a very advanced mathematical concept, more general than rectifiability, and in doing so encouraged, as the present book aims to do, a re-examination of the bases of many physical concepts that had been assumed to be differentiable. Curves of this type partition the plane into open sub-spaces; the main motivation of the

Figure 1.5 Construction of the Cantor set.

analysis given here is to show how to treat the physical phenomena that occur in such spaces.

A characteristic of the construction of the von Koch is that the number of arcs at stage n in the iteration is greater than that at stage $n-1$; but it could be otherwise, and an example of a different behaviour is provided by the Cantor set, shown in Figure 1.5.

In this case the number of arcs in the generator is less than the inverse of the scale, in fact $N = 2$ whilst $\eta = 1/3$, giving $\Delta = \log 2/\log 3 = 0.63 \ldots < 1$. The set is almost completely empty; it consists of a "dust" of singularities linked by a self-similarity relation; and here again, just as with the von Koch curve, any part is similar to the whole. The procedure is easily generalised and a topology can be associated with any value of the dimension.

1.3 What is a measure?

It can be seen in the construction of the von Koch curve by iteration that if η is a constant arc length and $\mathscr{P}$ is a finite sequence, then there is necessarily a relation between p and η; this will be the subject of the next chapter. Thus since $\lambda(p)$ is a length that depends on the parameter η, it can be replaced by $\lambda(\eta)$, which we write λ_η. Consider now the length λ_η of the approximation to the fractal curve by means of the gauge of length η; from 1.1 and 1.3 we have

$$\lambda_\eta = N_\eta \cdot \eta = l_0{}^\Delta [\eta]^{1-\Delta} \sim \eta^{1-\Delta}$$

showing that if $\Delta > 1$ the length tends to infinity as η tends to zero. This simple argument explains, for example, what Richardson showed, that the measure of a geographical coastline may be devoid of physical meaning.

The result is of major practical importance, for the dimension can thus be regarded as a value γ that divides the set of real numbers into two sub-sets according to the "measure" of the polygonal line $\lambda_{\eta,\gamma}$. For example, starting from the relation

$$\lambda_{\eta,\gamma} = N_\eta \cdot \eta^\gamma \tag{1.4}$$

how can we recover the fractal dimension of the curve? The solution is:

$$\lim_{\eta \to 0} [\lambda_{\eta,\gamma}] = \begin{cases} 0 & \text{for} \quad \gamma > \Delta(a) \\ \text{const} & \gamma = \Delta(b) \\ \infty & \gamma = \Delta(c) \end{cases} \tag{1.5}$$

What this states is that there is a value Δ of γ such that as the gauge length η is reduced the length $\lambda_{\eta,\Delta}$ approaches a finite limit. We can thus say that a curve Γ which has neither length nor area in the standard sense of these terms – $L(\Gamma) = \infty$ for $\gamma = 1$, $A(\Gamma) = 0$ for $\gamma = 2$ – has nevertheless a "content" $\lambda_{\eta,\Delta}$, that is a finite measure. The operation of finding the dimension Δ of the space is called the operation of measuring that space.

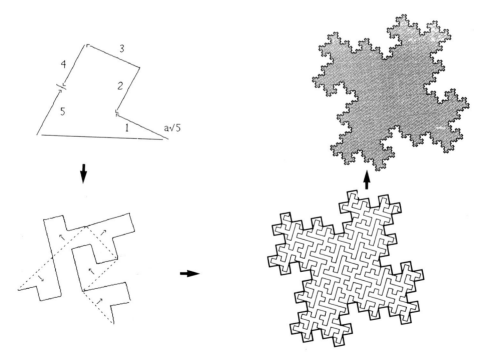

Figure 1.6 Construction of a space of dimension 2 by means of a broken line.

The striking simplicity of the process of constructing fractal curves by iteration and the possibility of having random objects with properties of self-similarity suggest that the values of the related parameters N, η can be chosen in an unlimited number of ways in order to produce a given value for the dimension Δ; why not then take N = 5, $\eta = 1/\sqrt{5}$ to give $\Delta = 2$? Nothing could be simpler, as is shown in Figure 1.6. The construction shows the progressive filling of the plane until an open surface is completely covered. This answers the question raised previously concerning covering a space of dimension 2 by a surveying process, that is, by means of arcs.

The bijection between a curve and a plane was first shown by Peano in 1890. But although the process for achieving this is identical with the construction of the von Koch curve there is a difference of scale between the two types of fractal limit: for $\Delta = 2$, in the limit $\eta = 0$ the construction leads to an infinity of pseudo-double points generated by the "final" approximation, and in the limit the curve cannot be used to define a partitioning of the plane.

1.4 From line to surface. A simple expression of fractality: self-similarity

The methods considered so far for constructing fractal objects are all based on the iteration of generator elements consisting of arcs arranged according to some particular topology. We shall now attempt to generalise this approach. Consider a space of dimension 1, that is, a line; this does not have the property of invariance under translation for after a translation it no longer consists of the points that constituted its original state – unless all points are considered to be identical. On the other hand, a change in length brought about by a scale function η, in particular by a factor less than 1, results in a segment which is in many respects very similar to the original line. A suitable choice of η will make it possible to cover the original line completely with N small pieces obtained by reducing the scale; we say that the line is self-similar to itself with the scale factor η such that $\eta(N) \sim N^{-1/3}$.

Correspondingly, the same procedure can be applied to a plane rectangular area, this time using small tiles giving a scale factor $(1/N)^{1/2}$; and for a volume the factor is $(1/N)^{1/3}$. The result is easily generalised to the case of a space of dimension Δ, for which the scale factor is

$$\eta(N) = (1/N)^{1/\Delta} \tag{1.6}$$

The formulation just given is identical to that given at the start, for finding the dimension of a curve by measuring arc lengths; and this suggests that there is a relation between the concept of dimension derived in paragraph 1.2, starting from a broken line, and that of self-similarity derived above. For this reason we say that if the above process is applied then Δ is the self-similarity dimension.

Application of the tiling procedure to the von Koch curve gives the self-similarity dimension $\Delta = \log 4/\log 3 = 1.2618...$ The result is easily generalised, giving

$$\Delta = -\log(N(\eta))/\log(\eta) \tag{1.7}$$

This form of the definition enables us to construct many objects by iteration of a generator of any type applied to a base structure also of any type. Figure 1.7 is an example.

Figure 1.7 The Sierpinski sieve. The sieve is constructed by applying three identical equilateral triangles to a basic equilateral triangle, the ratio of the side of the applied triangle to that of the base being $\eta = 1/2$. It can be shown that at the k'th step in the iteration the "gauge" number is $N_k = 3^k$ and that the gauge ratio is $\eta_k = (1/2)^k$. Assuming that the Mandelbrot relation $N_k \cdot \eta_k^{\Delta} = $ constant holds for all values of k, it follows that $\Delta = \log 3/\log 2 = 1.585...$

We now go more deeply into these scale properties. A rectifiable curve of length L can be measured by means of a gauge η and

$$\lambda_\eta = N_\eta \cdot \eta \to L \quad \text{as} \quad \eta \to 0 \tag{1.8}$$

For an area A the gauge is η^2:

$$A_\eta = N_\eta \cdot \eta^2 \to A \quad \text{as} \quad \eta \to 0 \tag{1.9}$$

and for a volume V in a space of dimension d the result is

$$V_\eta = V_\eta \cdot \eta^d \to V \quad \text{as} \quad \eta \to 0 \tag{1.10}$$

Suppose we approximate to a self-similar object of fractal dimension Δ, for example a polygonal curve, on two different scales with gauges η, η/b. Then $\lambda_\eta \sim \eta^{1-\Delta}$, $\lambda_{\eta/b} \sim (\eta/b)^{1-\Delta} = (1/b)^{1-\Delta} \cdot \eta^{1-\Delta}$ whence

$$\lambda_\eta / \lambda_{\eta/b} \sim b^{1-\Delta}$$

Similarly for a surface

$$A_\eta \sim \eta^{2-\Delta}, \quad A_{\eta/b} \sim (1/b)^{2-\Delta} \cdot \eta^{2-\Delta}$$
$$A_\eta / A_{\eta/b} \sim b^{2-\Delta}$$

and for a self-similar object in a space of dimension d

$$V_\eta / V_{\eta/b} \sim b^{d-\Delta} \tag{1.11}$$

Equation 1.11 conceals a change that has been made to the analysis so far as concerns measuring by means of the length of an arc. For a 2-dimensional space (d = 2) for example, we have used the relation

$$A_\eta = N_\eta \cdot \eta^2$$

while retaining the previous relation $N \sim \eta^{-\Delta}$, and so obtaining $A \sim \eta^{2-\Delta}$. This means that here we have used a gauge η^2 for measuring A, and this has the (physical) dimension of surface, not line. This change of gauge implies a "thickening" of the line; it can be taken into account by expressing the area as a thickened form of the polygonal line previously used to approximate a curve and measured by arc length:

$$A_\eta = \lambda_\eta \cdot \eta = N\eta^2 \sim \eta^{2-\Delta}$$

Whilst it is easy to understand this shift in the analysis, the reader should keep it well in mind when studying those physical properties that arise from fractality. He will find in the scientific literature, for example, scaling laws in which the exponent is a function of Δ in some cases and of $d - \Delta$ (i.e. $2 - \Delta$ in 2D, $3 - \Delta$ in 3D) in others. The reason for the difference lies in the type of measure that has been used in studying the particular physical phenomenon. We lay great stress on this point because much confusion can result from a failure to distinguish clearly enough between state variables related to a fractal interface, and therefore to Δ, and environmental variables related to the structure that supports this, and therefore to $d - \Delta$.

1.5 Fractal closed curves (loops): perimeter, area, density; fractal mass dimension; the concept of co-dimension

For an example of the difficulties that can arise here, consider an equilateral triangle, to each side of which we apply the procedure used to generate the von Koch curve. Iteration of this gives the following:

order of iteration	gauge length	segments	perimeter	area
	η_k	N_k	λ_k	A_k
1	1	3	3	$\alpha\,(=\sqrt{3}/4)$
2	1/3	12	4	$4\alpha/3$
k	$(1/3)\eta_{k-1}$ $= (1/3)^{k-1}$	$4N_{k-1}$ 3.4^{k-1}	$N_{k-1}\eta_{k-1}$ $3.(4/3)^{k-1}$	$A_{k-1}+$ $(1/4)(\eta_{k-1})^2 N_{k-1}$

Table 1.1 shows the first three steps in the iteration

It is obvious that as the iteration proceeds (k increases) the gauge length tends to zero and the perimeter to infinity; and it is easy to show that the area tends to the finite limit $(8/5)\alpha$. Thus we have a set of finite "content" bounded by a fractal curve of infinite length: this is the case for any object whose boundary is

Figure 1.8 Construction of the von Koch snowflake.

closed fractal curve. There seems something unnatural about the content of such an object: it is finite but it is not closed in the normal sense of the term. The boundary curve, although closed, defines an open set. Mandelbrot has shown that area and perimeter are related by the equation

$$\lambda_\eta \sim \eta^{1-\Delta}(A/\eta)^{\Delta/2} \quad \text{as} \quad \eta \to 0 \tag{1.12}$$

An example in which these ideas are relevant is the problem of determining the relation between current and magnetic flux in a fractal loop of von Koch type This is a problem that is not treated in the scientific literature, although it is of great practical importance. The apparently simple question attacks the very foundations of electromagnetic theory, that is, Ampère's theorem and its mathematical basis. The practical importance of the question will be seen from the discussion in Chapter 8, of electromagnetic phenomena in a fractal medium; it is the subject of a further work by the present author.

There are many objects of this type, characterised by very large perimeters but very small masses. There is a fractal dimension that is useful if one is interested in the content rather than the boundary, called the *fractal mass dimension*, writen Δ_M, and is not the same as the dimension Δ of the boundary. We can illustrate this by considering a self-similar object such as an aggregate, discussed in para. 6.3 and illustrated in Plate 9. The outline could be measured, but it could be useful to determine the law that gives the distribution of the mass. One method of doing this is to find the relation between the mass and the radius R of a "ball" used as a measure, where a "ball" is to be understood as a circle in a 2-dimensional space, a sphere in 3 dimensions and a hypersphere in higher dimensions. The scale laws that govern the fractal character here give the relation

$$M \sim R^{\Delta_M} \tag{1.13}$$

It is convenient at this stage to enquire into the relation between a fractal object considered as a whole and the dimension of the supporting environment in which it can be represented – for example, drawn on a sheet of paper. This leads to a concept, that of *co-dimension*, whose physical depth we shall see later.

Suppose a fractal object of dimension Δ is embedded in a space of dimension d. With R and Δ_M defined as above, the mass within a ball of radius R centred on the object is given by equation 1.13. But the ball also encloses a content C_H of the supporting medium given by

$$C_H \sim R^d$$

so the density of the fractal object relative to the support is

$$M/C_H \sim R - \zeta \quad \text{where} \quad \zeta = d - \Delta_M \tag{1.13a}$$

ζ is called the *co-dimension* of the fractal object. It is associated with the intensive properties of the object, such as its density, and becomes relevant when one needs to take note of the environmental variables of the fractal interface. The procedure we have given is typical of dimensional analysis extended to non-integral values for the dimension: it was by means of analysis of this type that Kolmogorov was able to establish the theoretical scaling laws for the development of turbulence.

1.6 Scaling laws with variable ratios

So far we have, to a certain extent, constrained the system to conform to simple self-similarity laws; how then can we attack the problem of a Cantor distribution in which the initial lengths of the two extreme segments are unequal? Consider Figure 1.10, in which the lengths of the left and right segments are 1/4 and 2/5 respectively of the original unbroken line: what is the fractal dimension of the set produced by the iteration? – can a dimension, in fact, be defined in this case?

Before attacking this problem we must attempt to define the concept of measure rather more precisely, using this in the sense of a *positive and additive set function*. We say that a measure is a function μ which for a given partition of a metric space into sub-spaces A, B assigns a positive number to each set such that

$$\mu(A \cup B) = \mu(A) + \mu(B)$$

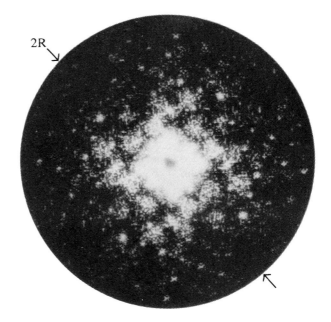

Figure 1.9 Method for measuring the fractal mass dimension of an aggregate. The photograph is of a fractal diffraction pattern taken in the focal plane of a lens, that is, the plane of the Fourier transform of the fractal object shown; and the "mass" here is the luminous flux. There is a relation between the fractal dimension of the object in the transform plane and that of the diffracting object (see para. 6.2).

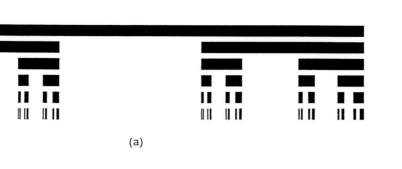

(a)

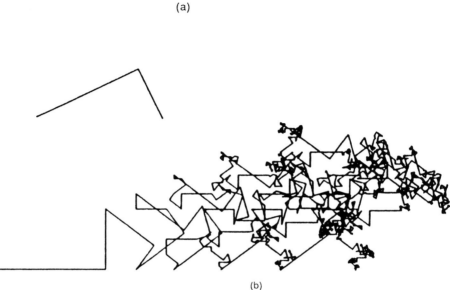

(b)

Figure 1.10 (a) Cantor set with unequal initial segments. (b) An example given by J. Feder of a fractal set of dimension 2.

and generally

$$\mu(\cup A_n) = \Sigma\mu(A_n) \tag{1.14}$$

The objective of a measure is to associate a set of values μ with sub-sets of the geometric space. We now apply this to the following set, which is distributed in a non-symmetric manner.

At the n'th iteration the number of segments is $N = 2^n$, with the shortest and longest having lengths $l_1^n = (1/4)^n$ and $l_2^n = (2/5)^n$ respectively.

The distribution of the lengths of the segments at this stage is given by the terms of the binomial expansion of $[(1/4)+(2/5)]^n$; so there are $C_k^n = n!/k! \cdot (n-k)!$ segments of length $l_1^k l_2^{n-k}$ with $k = 0.1.2,\ldots n$ and the measure, in the sense defined above, of the set created by this segmentation is

$$\mu = \Sigma_x C_k{}^n l_1{}^{k\Delta} \cdot l_2{}^{(n-k)\Delta}, \quad k = 0, 1, \ldots n$$
$$= (l_1{}^{\Delta} + l_2{}^{\Delta})^n$$

If the measure is to have any meaning it is important that its value remains finite and different from zero as the number of iterations (n) tends to infinity, in this case. Therefore we must have

$$l_1{}^{\Delta} + l_2{}^{\Delta} = 1$$

which leads to $\Delta = 0.611$.

The result is easily generalised: if the original is broken into a number of segments of fractional lengths l_i (with $\Sigma_i l_i < 1$) the fractal dimension Δ of the resulting generalised Cantor set is given by the equation

$$\Sigma_i l_i{}^{\Delta} = 1 \tag{1.15}$$

To illustrate this, and to show the open character of the concept of dimension, consider the case of a generator formed by two line segments of arbitrary lengths joined at a right angle, where at each iteration the lengths are reduced by factors l_1, l_2 which are different for the two arms, as in Figure 1.10b which, with $l_1 = 2/\sqrt{5}$ for the longer and $l_2 = 1/\sqrt{5}$ for the shorter, shows the start of the iteration; although not obvious from this, it follows from equation 1.1 that the dimension of the object thus created is $\Delta = 2$.

Figure 1.10b, taken from Feder, 1988 (see Bibliography), raises interesting questions. It is difficult to decide from the final figure the rule for its construction, that is, at what angle each new piece is to be joined to its predecessor – whether in fact there is a rule or whether the angle is chosen at random. In fact, whatever the case the dimension is always 2, although of course the form of the curve will be different for different choices. If there is a rule, so that the successive angles are determined, then there is an invariant that describes the construction as a whole, which we call the *spin* of the curve.

1.7 From self-similarity to self-affinness

The sets considered so far have been based on an assumption of self-similarity which is such that a (generalised) sphere remains a sphere under scale reduction.

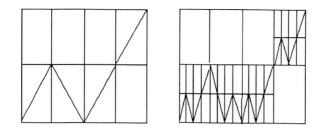

Figure 1.11 Example of a self-affine set.

But consider the set of Figure 1.11. The graph here is not self-similar because the scale factors are different for the two axes; but it is self-affine. We should recall here that affine geometry does not differentiate between a square and a general rectangle, or between a circle and a general ellipse, etc. Thus approximation by arc length, which requires an "isotropic" space, is not appropriate here and we have the question of what measure to use in connection with such graphs, how to devise an approach that is general and at the same time is relevant to the multiplicity of cases to which self-affinness gives rise. This is the aim of the next chapter.

1.8 Language, fractal geometry and recurrence series

We have said previously that fractal geometry is based on a simple principle: the iteration to infinity of a generator group:

generator + iteration → complex set (fractal)

A measure is constructed for the set to which this process finally leads, whose value depends on the generator.

It can be easily imagined that the generator can be any transformation group whatever; we shall return to this point, but first we consider a method for coding the information such that for the example of the von Koch curve (Fig. 1.12) we write

$$0 = \text{rotation through } \frac{\pi}{3}, \quad 1 = \text{rotation through } \frac{-2}{3}\pi$$

a   b

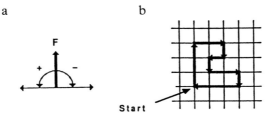

FFF-FF-F-F+F+FF-F-FFF

c

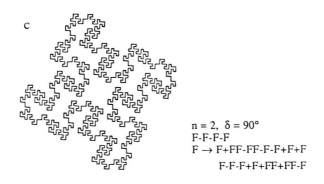

n = 2, δ = 90°
F-F-F-F
F → F+FF-FF-F-F+F+F
F-F-F+F+FF+FF-F

Figure 1.12 (a) Turtle interpretation of string symbols F, +, −. (b) Interpretation of a
string. The angle increment δ is equal to 90°. Initially the turtle faces up. (c)
Example of Koch curve generated using L-systems; quadratic von Koch
island.

The first two orders of the curve are then

n = 1   0   1   0   0

n = 2   0100 0101 0100 0100

The coding can also be written as a recurrence series based on the two
transformations

$$0 \rightarrow 0100, \quad 1 \rightarrow 0101$$

so that at stage n the resulting word has 4^n letters (0, 1). Note also that the object
obtained is supported by the Cantor set which in turn can be written

$$0 \rightarrow 000, \quad 1 \rightarrow 100$$

where 1 means "existence" and 0 means "non-existence". Here the word at the n'th stage has 3^n letters. M. Dekking has used this construction to generate random Cantorian sets in the plane.

These codes can be given different properties; for example, reflexion in the Thue-Morse series for which $0 \to 01$, $1 \to 10$. This has some very specific properties, for example if ω is a word and ω' is the same word in which the letters have been interchanged (the mirror effect) then $\tau(\omega') = \tau'(\omega)$. One can imagine that it should be possible to deduce many theorems directly from the form or the generator, just as it is simple to give them the geometrical representation.

Returning to the von Koch curve, there will be conditional transition probabilities p(01), p(00), p(10). We can assume that there will be an angle $\varphi = \Sigma p_i \theta_i$ which will be closely related to the fractal dimension Δ and to the phase factor which appears in the analysis of the concept of time in fractal geometry; but this remains to be proved.

The language can of course be extended to more complex geometrical objects. For example, in 1974 Penrose constructed a non-periodic tiling of the plane characterised by a pentagonal symmetry, coded as $A \to AAB$, $B \to AB$. This is shown in Figure 1.13.

Many fractals can be thought of as sequences of primitive elements. To produce fractal strings generated by a language we must use a geometrical code and here we describe one based on the Logo "turtle" (H. Abelson & A.A. Disessa), originally proposed by Szilard & Quinton. A state of the turtle is defined by the triplet (x, y, α) where (x, y) are the cartesian co-ordinates of its position and α is the direction in which it is facing – its "heading". Given a step length d and an angular increment θ the turtle's motion is defined by $(x, y, \alpha, d, \theta)$.

If F means "1 step forward", $\theta = 90$ and $+/-$ means "turn through $+90/-90$", Figure 1.12 shows the fourth approximation to the quadratic von Koch island:

$$\omega: F + F + F + F \quad \text{and} \quad G: F + F + F - F - FF + F + F - F$$

where ω is the initial stage and G is the transformation group.

It will be clear from the above that in these constructions it is the angles that are conserved; it is these that convey the main properties of the fractals, in particular those that are related to the irreversibility of time. Points to note are –

1. there is an invariable order in the iteration which must never be inverted
2. it is through the rotations that the progress of the turtle over the geometrical structure is related to the environment, that is, the supporting space in which the path is traced. At the end of Chapter 2 we shall return to the importance of

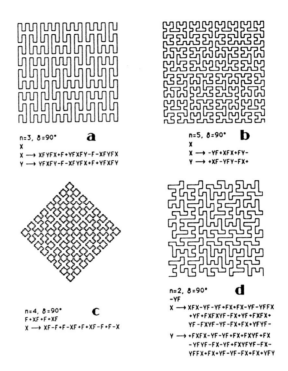

n=3, δ=90°   **a**
X
X ⟶ XFYFX+F+YFXFY-F-XFYFX
Y ⟶ YFXFY-F-XFYFX+F+YFXFY

n=5, δ=90°   **b**
X
X ⟶ -YF+XFX+FY-
Y ⟶ +XF-YFY-FX+

n=4, δ=90°   **c**
F+XF+F+XF
X ⟶ XF-F+F-XF+F+XF-F+F-X

n=2, δ=90°   **d**
-YF
X ⟶ XFX-YF-YF+FX+FX-YF-YFFX
+YF+FXFXYF-FX+YF+FXFX+
YF-FXYF-YF-FX+FX+YFYF-

Y ⟶ +FXFX-YF-YF+FX+FXYF+FX
-YFYF-FX-YF+FXYFYF-FX-
YFFX+FX+YF-YF-FX+FX+YFY

Figure 1.13 "Classic" space-filling curves and the corresponding L-systems. (a) Peano [1890] curve, (b) Hillert [1891] curve, (c) A square-grid approximation of the Sierpiński [1912] curve, (d) Quadratic Gosper curve [Dekking 1982].

the conservation of the angles in the irreversibility factors that control the progress of the turtle.

A turning point is a singular point which, apart from the information contained in its co-ordinates, condenses information concerning the environment (Fig. 1.12). Fractal geometry enables us to inerpret the singularity as a local concentration of global information; the link between time irreversibility and the fractal nature of the geometry is the expression of this condensation.

Chapter 2

Measures of dimension; time in fractal geometry

2.1 Practical methods; different measures of dimension

So far our study, based as it is on measurement by arc length, has taken no account of physical reality. The concept of co-dimension has brought this in, certainly, but only to a limited extent; although the approach taken offers rich prospects and perspectives, it may seem incomplete and in need of further development. Further, this deliberate neglect of the physical reality may seem rather suspect from the point of view of the physical processes that we shall be considering.

To repair the omission we take an empirical approach and consider the physical situation in which the length λ_η is the *transfer function* for a possibly irreversible process, defined as follows:

TRANSFER FUNCTION. *Let φ, J be the input and output variables respectively of a physical process, and suppose that in the space of frequencies* p *there is a parameter $\lambda(p)$ and a relation* $J(p) = \lambda(p)\varphi(p)$. *Then, whatever the form of the function $\lambda(p)$, we say that $\lambda(p)$ is the linear transfer function between φ and* J.

The mathematical analysis is based on the process of covering the fractal object with a sequence of (generalised) spheres of dimension equal to that of the space in which the object is embedded: thus we construct a measure for a plane curve (i.e. a curve in 2-dimensional space) by covering it with discs of diameter η and therefore area proportional to η^2 (Fig. 2.1).

This gives rise to various questions, for example: do all the discs have to be of the same size? is there a relation between this measure and that based on arc lengths ("yardsticks")? are these measures in any sense identical? and so on.

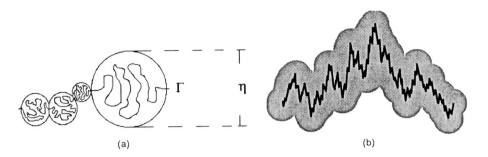

Figure 2.1 (a) Hausdorff measure, characterised by the distribution of the sizes of the discs. (b) Minkowski "sausage" for a graph of Brownian movement.

The motivation for the analysis of paragraphs 2.3, below, and its successors, was the desire to establish a sound mathematical basis for the physical models TEISI (Transfer of Energy at a Fractal Interface, 1979) and TEISA (Transfer of Energy at a Self-Affine Interface, 1988), which were devised with the aim of linking fractal dimension and entropy. This is the outcome of a long study by C. Tricot and the present author, undertaken with the aim of bringing generality into what seemed, up to the end of the 1970s, to be a number of special hypotheses.

2.1.1 *The Hausdorff dimension*

As a premiminary to constructing a rigorous definition of dimension, consider the way in which Lebesgue defines the "area" of a set Γ of points in the plane. For a given $\eta > 0$ he considers the covering of Γ with discs D_j of radius $r_j < \eta$; the sum of the areas of these discs will depend on the distribution of the radii r_j, and will have a lower bound, $\mu_\eta(\Gamma)$ say:

$$\mu_\eta(\Gamma) = \inf(\Sigma \pi r_j^2) \tag{2.1}$$

The Lebesgue external measure is then defined as

$$\mu(\Gamma) = \lim_{\eta \to 0} \mu_\eta(\Gamma) \tag{2.2}$$

In a space of dimension d a corresponding measure $\mu^{(d)}$ can be defined by enclosing the points in d-dimensional spheres.

For the case of the plane, the value 2 taken for the dimension of the covering discs is larger than it need be in certain cases, and a smaller measure can be obtained by taking r_j^α instead of r_j^2, where α is a parameter to be determined. This leads to the Hausdorff measure, which is suited to use with fractal curves:

$$\text{Let} \quad H_{\alpha,\eta} = \inf(\Sigma r_j^\alpha) \tag{2.3}$$

where $\Gamma \in \{\cup D_j\}$, $r_j < \eta$ and α, η are parameters whose values are at our choice. The Hausdorff measure is then

$$H_\alpha(\Gamma) = \lim_{\eta \to 0} H_{\alpha,\eta}(\Gamma) \tag{2.4}$$

If the measure is made in a space of dimension d and if $\alpha = d$ then the Hausdorff measure is the same as the Lebesgue, to within a numerical factor. For general α, H_α is an α-dimensional measure since, using the results of Chapter 1, if we construct a set Γ' homothetic with Γ with ratio b then from § 1.4

$$H_\alpha(\Gamma) = b^\alpha H_\alpha(\Gamma) \tag{2.5}$$

The Hausdorff measure has the property that for all Γ there is a critical value α_0 such that

$$\alpha < \alpha_0 \to H_\alpha(\Gamma) = \infty$$
$$\alpha > \alpha_0 \to H_\alpha(\Gamma) = 0 \tag{2.6}$$

Thus we can define the smallest α such that H is zero or the largest such that H is infinite:

Measure	$H_\alpha(\Gamma) = \infty$	$H_\alpha(\Gamma)$ undefined	$H_\alpha(\Gamma) = 0$
Dimension	$\alpha < \alpha_0$	$\alpha = \alpha_0$	$\alpha > \alpha_0$

This (Dedekind) cut in the real numbers defines $\alpha_0 = \dim(\Gamma)$, the Hausdorff dimension of Γ.

2.1.2 *The Minkowski-Bouligand dimension*

Let $N_\eta(\Gamma)$ be the minimum number of spheres of radius η (all therefore identical) needed to cover the set Γ; the generalised volume of this set of spheres is $\eta^\alpha N_\eta(\Gamma)$, to within a numerical factor as usual, and

$$\eta^\alpha N_\eta(\Gamma) \geqslant H_{\alpha,\eta}(\Gamma) \tag{2.7}$$

Then for all values of α such that $H_\alpha = \infty$ (i.e. $\alpha > \alpha_0$)

$$\lim_{\eta \to 0} \eta^\alpha N_\eta(\Gamma) = \infty$$

Taking logarithms

$$\lim_{\eta \to 0} [\alpha \log(\eta) + \log N\eta(\Gamma)] = \infty \tag{2.8}$$

and dividing by $\log(1/\eta) = -\log(\eta)$

$$\lim_{\eta \to 0} [\{\log N_\eta(\Gamma)/\log(1/\eta)\} - \alpha] \geqslant 0$$

Writing now (for reasons that will appear shortly)

$$\lim_{\eta \to 0} [\log N_\eta(\Gamma)/\log(1/\eta)] = \Delta_{MB}(\Gamma)$$

we have

$$\Delta_{MB}(\Gamma) \geqslant \alpha \tag{2.9}$$

i.e.

$$\dim(\Gamma) \leqslant \Delta_{MB}(\Gamma) \tag{2.10}$$

This new measure Δ_{MB} is called the *Minkowski-Bouligand dimension*, or, for obvious reasons, the *logarithmic density* of the set Γ.

It might seem that this measure contradicts some of the definitions of a dimension given in Chapter 1; to investigate this, we consider a curve or a set Γ in a space of 2 dimensions, that is, in the plane of the paper, and define the

"Minkowski set" as the union of the set of discs of radius η centered on the points of Γ, $\cup D_\eta(\Gamma)$ – what is called colloquially the *Minkowski sausage*, see Figure 2.1b. Let $N(\eta)$ be the minimum number of discs needed to cover Γ and $A_2(\eta)$ be area of this set of discs; from the analysis of Chapter 2 it is seen that

$$A_2(\eta) \sim N(\eta)\eta^2 \tag{2.11}$$

But from the definition of Δ_{MB} we have

$$N(\eta)^{\Delta_{MB}} \sim 1 \tag{2.12}$$

and hence

$$A_2(\eta) \sim \eta^{2-\Delta_{MB}} \tag{2.13}$$

Thus we have an alternative definition of Δ_{MB}:

$$\Delta_{MB} = \lim_{\eta \to 0} [2 - \{\log A_2(\eta)/\log \eta\}] \tag{2.14}$$

If we compare equation 2.13 with the expression for the length of a fractal curve of dimension Δ, $\lambda_\eta \sim \eta^{1-\Delta}$, we see that it is equivalent to writing $A_2(\eta) \sim \eta \cdot \lambda_\eta$, with Δ replaced by Δ_{MB}. So we can say that the Minkowski sausage is obtained by drawing the curve with a pen of thickness η. Further, we have from 2.13 $\log A_2(\eta)/\log \eta \sim 2 - \Delta_{MB} = \zeta$, the co-dimension.

In the above we have taken all the discs used to cover the curve as having the same measure; when we come to consider the concept of *multifractality* in Chapter 4 we shall see that this need not be the case.

The definition of Δ_{MB} is easily generalised to a space of dimension d:

$$\Delta_{MB} = \lim_{\eta \to 0} [d - \{\log A_d(\eta)/\log \eta\}] \tag{2.15}$$

where $A_d(\eta)$ is the volume of the d-dimensional "sausage" formed by the union of the covering set of d-dimensional spheres of radius η.

This is a suitable place to remark that since the measure concerns an approximation $\lambda_\eta(\Gamma)$ to the length Γ obtained by means of a gauge of length η, the fractal dimension can be found experimentally by plotting (on log-log paper) $\log \lambda_\eta(\Gamma)$ against $\log \eta$ and fitting the best straight line: the slope of this is $1 - \Delta$. Alternatively the approximation can be done by covering the curve with small discs, as above, when the slope of the regression line of $\log A_2(\eta)$ against $\log \eta$ will be $2 - \Delta_{MB}$. The two methods will not necessarily give the same value for the dimension, except in certain particular cases: in the case of self-similarity Δ_{MB} is exactly equal to Δ, but this is one of the particular cases. There are in fact other dimensions, including what is called the *packing dimension*, which we study next.

2.1.3 *The packing dimension*

This is due to C. Tricot. As before let Γ be a curve in the plane and D_j a set of discs of radii $r_j \leqslant \eta$ with centres on Γ, but now such that *no two overlap*. We are interested in the sum Σr_j^α and define

$$T_{\alpha,\eta}(\Gamma) = \sup \Sigma r_j^\alpha \tag{2.16}$$

and

$$T_\alpha(\Gamma) = \lim_{\eta \to 0} T_{\alpha,\eta}(\Gamma) \tag{2.17}$$

As in deriving the Hausdorff measure, we can show that there is a value α_0 defined by a cut in the real numbers, such that

$$\alpha < \alpha_0 \to T_\alpha = \infty$$
$$\alpha > \alpha_0 \to T_\alpha = 0 \tag{2.18}$$

where it is important to note that α_0 is precisely Δ_{MB}. Finally, we arrive at definitions of the packing measure and packing dimension of Γ as follows.

$$\text{Let} \quad T'_\alpha(\Gamma) = \inf \Sigma T_\alpha(\Gamma_n) \tag{2.19}$$

where the lower bound is taken over all the partitions of Γ into sub-sets Γ_n.

There is a critical value of α defined by a cut in the real numbers as in equation 2.18; we say that this value is the *packing dimension* of Γ, written $\text{Dim } \Gamma$ (the capital D distinguishing this from $\dim \Gamma$) and that the corresponding value of $T'_\alpha(\Gamma)$ is the *packing measure*. There is this relation:

$$\dim(\Gamma) \leqslant \text{Dim}(\Gamma) \leqslant \Delta_{MB}(\Gamma) \tag{2.20}$$

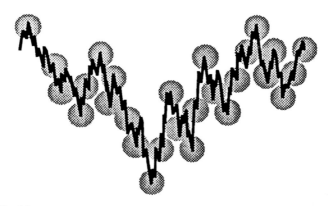

Figure 2.2 Packing measure.

A set is said to be (*very*) *regular* if $\dim(\Gamma) = \text{Dim}(\Gamma) = \Delta_{MB}$, that is, if the Hausdorff, Minkowski-Bouligand and packing dimensions are all equal. A very important property of such a set is that if Γ is regular, then for any set F

$$\dim(\Gamma \times F) = \dim \Gamma + \dim F \tag{2.21}$$

This has consequences for the co-dimension of $\Gamma \times F$. For if, for example, $\dim \Gamma = \Delta$ and $\dim F = 1$, $\dim(\Gamma \times F) = \Delta + 1 = D$, say; the co-dimension ζ of Γ is $2 - \Delta$ (since Γ is a curve in the 2-dimensional plane), which can be written $\zeta = (2+1) - (\Delta + 1) = 3 - D$. The relation $D = \Delta + d$ holds for all values d of $\dim F$, hence the co-dimension of $\Gamma \times F$ is independent of the space in which the regular fractal set Γ is embedded, provided this of dimension greater than 1. Conversely, any independence of a dimension of, for example, the Euclidean dimension of the space suggests that this could act as a co-dimension for regular sets.

2.2 Two methods for measuring the fractal dimension

Whilst the method just described for measuring fractal dimensions lends itself readily to computation, for example when the fractal set is examined with an image analyser, it is unfortunately very imprecise, for only rarely, contrary to what is often stated, does the log/log plot give a well-defined straight line that can be relied upon. In a sense, the Minkowski set is too "thick" to enable the fitting to be done with any accuracy. A better method is what is called the "boxes" method (although a better name would have been "mesh method").

2.2.1 *The boxes method in R²*

The basis of this method is the covering of the set Γ with a sequence of square meshes of decreasing side η_n and counting the number N_n of squares that contain at least one point of Γ: this is illustrated in Figure 2.3.

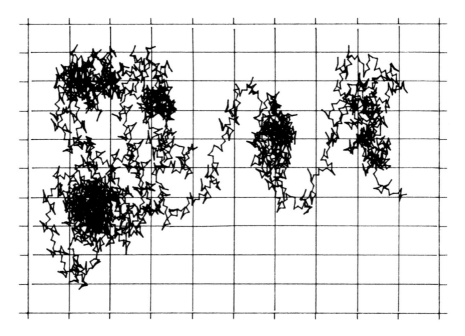

Figure 2.3 Boxes method for finding the Minkowski-Bouligand dimension. Here $1/\eta$ is not an integer and η is strictly between 0 and 1.

The dimension is then

$$\Delta_{MB} = \lim_{\eta_n \to 0} [(\log N_n)/(\log 1/\eta_n)] \tag{2.22}$$

(where $N_n \to \infty$ as $\eta_n \to 0$).

If $\log N_n$ is plotted against $\log 1/\eta_n$ the slope of the line gives Δ_{MB}.

But although very easy to use, this method too has serious disadvantages. If $1/\eta_n$ is not an integer the squares of side η_n will usually extend beyond the graph of Γ on the left and on the right, thus falsifying the count and introducing irregularities into the log/log plot, especially when $1/\eta_n$ is large. Thus if the graph is defined over the interval $[0, 1]$ a convenient way to ensure that the projection on Ox of any square containing points of the graph is included in $(0, 1)$ is to develop the mesh by continued bisection, so that $\eta_n \sim 2^{-n}$. However, η_n then decreases rapidly and one soon gets down to the precision of the data, when the process must stop; and only a few points are obtained for the log/log plot, insufficient to give a reliable result.

Non-binary sequences can be used which have other disadvantages and there are methods for correcting these and also for reducing the statistical errors; but not all the inaccuracies can be corrected, for these are inherent in such simple methods: for example, those due to the fact that since N_n takes only integer values it changes discontinuously when the mesh size changes from η_n to η_{n+1}.

The methods so far described have depended on the covering of the curve Γ with sequences of identical circles or squares of characteristic dimension η_n; we could instead use other geometrical shapes, still keeping this characteristic dimension. However, measures based on sets of the type of the Minkowski "sausage" require that the reductions in the scale of η do not depend on the orientation of the covering elements; so it is easy to see that a measure of the Minkowski-Bouligand type is not well adapted to a self-affine function, as for example Figure 1.11, for which the scaling properties are different in the directions of the two axes. We therefore need a method that is better suited to such a case, and we now consider what is called the *variations method*.

Let f be a continuous function of the single variable x and Γ its graph, that is, the set of points with co-ordinates $[x, f(x)]$. Let $S(x, \eta)$ be a horizontal segment of length η centred on the point $[x, f(x)]$ of the curve, as in Figure 2.4: this is

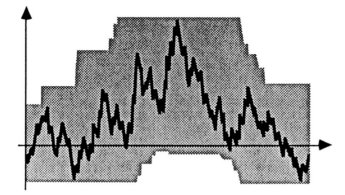

Figure 2.4 Basis of the variations method for finding the fractal dimension of a self-affine graph (C. Tricot).

called the *horizontal structuring segment*. As S traverses the graph it sweeps an area $A_2(\eta)$, the union of all the $S(x, \eta)$, and it can be shown that, assuming that $f(x)$ is not a constant,

$$\Delta_{\text{MB}} = \lim_{\eta \to 0} [2 - \log A_2(\eta)/\ln \eta] \tag{2.23}$$

This is the basis of the method.

2.2.2 Variations method

If Γ is a self-affine fractal curve defined in the interval $(0, 1)$ of x there will be at least one part of this interval in which $f(x)$ is nowhere, or almost nowhere, derivable (a term meaning the same as differentiable, used here for the first time in this book and which we shall need to use later), because every point in this sub-interval is a singular point. If $p(x, x')$ is the slope of the line joining $[x, f(x)]$ and $[x', f(x')]$, it follows from the concept of fractality that the upper bound of $|p(x, x')|$ as $x' \to x$ is infinite. *It is of course the way in which this upper bound approaches infinity that determines the fractal dimension*, a property we shall meet when studying problems in physics. The mathematical task, however, is to assign a measure to this behaviour, which is typical of non-rectifiable curves, and to achieve this we define a function $v(x, \eta)$ as follows:

$$v(x, \eta) = \sup [f(x')] - \inf [f(x')], |x' - x| \leqslant \eta \tag{2.24}$$

where x' is in the interval of definition of $f(x)$, here $[0, 1]$. $v(x, \eta)$ is called the η-oscillation of $f(x)$ about x. Then

$$V(\eta) = \int_0^1 v(x, \eta)\, dx \tag{2.25}$$

is called the η-variation of $f(x)$ and tends to 0 as $\eta \to 0$. One might conjecture that the rate of increase of $V(\eta)$ as η increases is directly related to the fractal dimension of the graph.

It can be shown, using a theorem of Tricot and Roques-Carmes, that the area swept out by the horizontal structural segment is $A_2(\eta) = V(\eta)$; and from this and equation 2.23 we have

$$\Delta_{\text{MB}} = \lim_{\eta \to 0} [2 - \log V(\eta)/\ln \eta] \tag{2.26}$$

It is easily seen that, in contrast to the previous methods, this is practically independent of a change of scale. Thus if η is halved so also are $v(x, \eta)$ and its integral $V(\eta)$, resulting simply in a shift in the log/log plot. The reader will find, by actually applying the method to self-affine curves, that it gives very well-defined regression lines.

A study of the brief analysis just given will show that image analysis, in spite of the effort expended on it by various mathematicians, is far from providing the simplest and most reliable method for investigating the fractal nature of an object. Provided that one knows which set will determine the result it can be much preferable to set up some physical interaction with this, from which to deduce the fractal dimension or dimensions. But this must be done with care, for there are pitfalls to be avoided, the commonest of which is to assign a characteristic length ξ to a physical process and to use this as the gauge for measuring the fractal by approximating with step length ξ: for example, in a problem in chemical diffusion setting $\xi \sim \sqrt{(D_{iff} \cdot t)}$, where D_{iff} is the diffusion coefficient and t the time. Such a procedure is questionable because, whilst keeping to the mathematics, it ignores the parameterisation that underlies the form and thus the measure of the fractal curve or graph. This will be taken up in the next chapter.

2.3 Relation between time and measure: parameterisation of fractal curves

We now take up the question of the time involved in a fractal measure: the reader will have learned how a fractal dimension can be measured, but will not know how much time is required to carry out the process. The more restricted problem that we now attempt to solve is to find the relation between the length of a fractal curve and the time required to measure this.

Mathematically expressed, the curves Γ that we have been studying have no double points and are continuous injective mappings γ of a segment $[a, b]$ on to some other segment, so that to every point of $[a, b]$ there corresponds a point of Γ: this is equivalent to an elastic deformation. Thus defined, Γ is a parametrised curve, and the parametrisation γ is one way of defining Γ. For example, the graph of a function $f(x)$ in cartesian co-ordinates is the set of points $[x, f(x)]$, where x runs through the set over which $f(x)$ is defined, and here the parameter is x.

The von Koch curve is parametrised, since at each stage it is defined by the rule for constructing the segments: thus

Stage 1: $[0, 1/4], [1/4, 1/2], [1/2, 3/4], [3/4, 1]$

Stage n: $[k/4^n, (k+1)/4^n]$, $k = 0, 1, \ldots 4^n - 1]$

Every explicit geometrical construction, in fact, corresponds to some parametrisation, and indeed it could be said that in any such construction parametrisation is done unconsciously. But the situation is fundamentally different if the curve is defined implicitly, by an equation of the type

$$F(x, y) = 0 \qquad (2.27)$$

for such an equation does not necessarily define a unique curve: there may be several solutions, or even none. However, every one agrees that a curve is defined if functions $g(t)$, $h(t)$ of a variable t can be found such that

$$F[g(t), h(t)] = 0 \qquad (2.28)$$

and in this case t provides the parametrisation of the curve.

Physically, a continuous curve can be visualised as the trajectory of some unspecified flying object, when the relevant parameter is *time*: the curve is defined as the set of positions of the object at each instant. However, *since the speed with which the object describes its trajectory is arbitrary, there are an unlimited number of possible parametrisations of any given curve.* Thus the aim of a research study is not simply to find a law relating the speed to the position on the trajectory, but to find the law that is most relevant to the ultimate purposes for which the research is being undertaken on the assumption that the fractal expresses some physical phenomenon.

To investigate how the parametrisation of a curve Γ can be related to its possible fractal dimension, consider again the concept of measure. In practice, any parametrisation leads to a measure; for example, if $A_1 A_2$ is an arc of Γ with $A_1 = \gamma(a_1)$, $A_2 = \gamma(a_2)$, then $A_1 A_2$ is the image of $[a_1, a_2]$ under γ and its measure is $(a_1 - a_2)$ – this could be, for example, the time taken by the object in traversing the arc $A_1 A_2$. Expressed very abstractly, the measure of a subset E of Γ is the length – or more correctly, the Borel measure – of the inverse image $\gamma^{-1}(E)$.

p-adic parametrisation of a fractal curve

One of the first attempts to give an intrinsic definition of a physical fractal was made by L. Nottale in 1983, who focussed his research on questions of infinities and non-differentiability. He considered first the case of a fractal curve in the complex plane, starting with a generator $\Gamma 1$ consisting of p segments each of length $1/q$ and end points $x = 0$, $x = 1$; and developed his analysis in terms of a p-adic representation of a parameter defined on the fractal, that is, a parameter $s = 0 \cdot s_1 s_2 \ldots s_k \ldots = \Sigma s_k p^{-k}$ with s_k one of the integers $0, 1, \ldots p - 1$.

Labelling the segments 0 to $p - 1$, their end-points in cartesian and in polar co-ordinates are

$$Z_j = x_j + i y_j = q^{-1} \exp(i\theta_j), \quad j = 1, 2, \ldots p \tag{2.29}$$

Alternatively we can use the polar angle ω_j of segment j or the angle α_j between $j - 1$ and j. There are the following relations:

$$Z_{j+1} - Z_j = q^{-1} \exp(i\alpha_j) \tag{2.30}$$

$$\omega_j = \sum_1^j \alpha_k \tag{2.31}$$

$$\sum_0^{p-1} q^{-1} \exp(i\omega_j) = 1 \tag{2.32}$$

We now define a parameter s on the fractal Γk, a normalised curvilinear co-ordinate, as a base-p fraction

$$s = 0 \cdot s_1 s_2 \ldots = \Sigma s_k p^k \tag{2.33}$$

where each s_k is one of the integers $0, 1, \ldots p - 1$. We can then write the fractal equation in the form

$$Z(s) = Z_{s1} + q^{-1} \exp(i\omega_{s1})[Z_{s2} + q^{-1} \exp(i\omega_{s2})[Z_{s3} + \ldots]] \tag{2.34}$$

and if

$$\varphi_{sk} = \omega_{s1} + \omega_{s2} + \ldots \omega_{sk-1} + \theta_{sk} \tag{2.35}$$

the parametric equation becomes

$$Z(s) = \Sigma \theta_{sk} q^{-k} \exp(i\varphi_{sk}) \tag{2.36}$$

In terms of s, $x(s)$ and $y(s)$ are fractal "functions" for which successive approximations can be constructed. Their curves have the same dimension $\log p / \log q$ as the original one.

The structure of the equation 2.36 is remarkable since it evidences the part played by the p-adic decomposition through p and q in the plane

$$Z(s) = \Sigma^\infty C_k(s) q^{-k} \quad \text{with} \quad s = \Sigma s_k p^{-k} \tag{2.37}$$

An intrinsic building of a fractal curve may also be made. Placing ourselves on Γ_n, we only need to know the orientation from the previous segment of length q^{-n}. On the generator Γ_1 these angles have been named α_j. The problem is now to find $\alpha(s)$. The parts of the Γ_n which are common with Γ are characterized by rational parameter s written with n figures in the counting base p, $s = 0 \cdot s_1 s_2 \dots s_n$. Let us designate by s_n the last non null figure of s, i.e.

$$s = s_1 p^{-1} + s_2 p^{-2} + \dots s_n p^{-n} \tag{2.38}$$

It is easy to verify provided that $\alpha_j = \omega_0 - \omega_{p-1}$ (which is a necessary condition for self avoidance), that the relative angle between segment number $sp^{-n} - 1$ and segment $s_n p^{-n}$ on Γ_n is given by $\alpha(s) = \alpha_{sn}$.

So the fractal may be defined in a very simple way uniquely from the $(p-1)$th structural angle and independently from any particular coordinate.

Let us now consider the difference between two successive terms.

$$s = \Sigma s_k p^{-k} \quad \text{and} \quad s + \eta = \Sigma(s_k + 1) p^{-k} \tag{2.39}$$

that is $(q^\Delta)^k = 0 \cdot 0_1 0_2 \dots 0_{k-1} 1_k = p^{-k}$ in the counting base p.

$$1/\eta = (q)^k = (p^k)^{1/\Delta} = (p)^{1/\Delta}$$
$$1/\eta = (p)^{1/\Delta} \tag{2.40}$$

where p is a generalised frequency. This equation gives the relationship between the frequency distribution (i.e. time) and the gauge of the measure (i.e. space). This result is the first expression of the TEISI model. *This relation is strictly intrinsic.* The codimension does not play any role in the definition of the parametrisation.

If p is written $p = i\omega$, and $\Delta = 1$, $\eta = 1/i\omega$ the gauge behaves like a 1-dimensional density *capacitor measure*. If $\Delta = 2$, $\eta = 1/\sqrt{i\omega}$ the gauge behaves like a 2-dimensional density *diffusive measure*. Both measures have a profound physical meaning, usually understood as specific, for instance dielectric relaxation and diffusive relaxation respectively. In fact, in the above interpretation, (i) these specific properties may be understood in the frame of a theory of measure (typically a macroscopic theory) in which a parametrisation of the space is added in place of any local interpretation, and (ii) their characteristics are no longer focused on specific values of Δ (1 or 2) but may be generalised to a real set (equation 2.40 where $p = i\omega$).

Another way of expressing this is to say that the flying object is nowhere stationary.

We now consider how the parametrisation can be used to measure length, so as to relate it explicitly to the fractal dimension. We shall find in particular that different parametrisations of the same object lead to different measures: this can raise tricky problems of communication when the same physical phenomenon is studied by different research teams using different types of measure.

Take for example the von Koch curve with its natural parametrisation: the image of each interval $[k \cdot 4^{-n}, (k+1) \cdot 4^{-n}]$ is a part $A_k A_{k+1}$ of the curve, of "diameter" 3^{-n}. The measure of $A_k A_{k+1}$ is 4^{-n}, that is, the projectile covers a distance 3^{-n} as the crow flies in a time 4^{-n}; and since this holds for all n, the speed along the curve – here $(4/3)^n$ – becomes infinite as the length becomes infinite. Thus so classical a concept as speed becomes inapplicable when one is forced to work with infinities, and consequently with non-standard numbers.

The practical result is that it is meaningless to speak of lengths and speeds in a fractal medium, and only measures of trajectories can be discussed meaningfully.

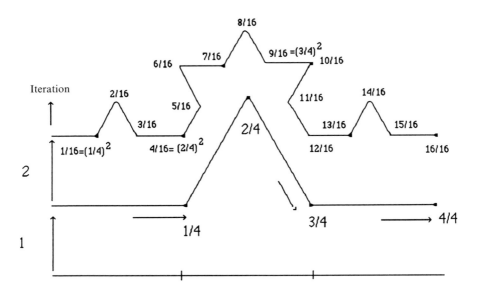

Figure 2.5 Parametrisation of the von Koch curve.

2.3.1 *Measure and length*

If the curve Γ is a straight-line segment its length is its "diameter"; if not, it can be approximated by summing the diameters of a set of linear segments of given length whose union gives the best approximation to the curve, and the length is found, as in Chapter 1, by steadily reducing the length of the approximating segment. We now consider how this can be expressed with precision in terms of the parametrisation of Γ.

By definition, we can associate with Γ a parametrisation γ defined on a certain interval $[a, b]$, or, what comes to the same thing, a support measure Γ. The natural approximation to Γ is a sequence of small arcs, all of the same measure t, which serve in a sense to smooth Γ; so for each z in $[a, b]$ we take an interval $I(z, t)$ of length t, containing z, the image of which under γ will be an arc of Γ of measure t. We must take care to ensure that the points chosen are not equidistant, as that would give special weighting to certain points. We have now that Γ is the union of $N = (b-a)/t$ arcs of mean diameter $\langle \xi \rangle$, where

$$\langle \xi \rangle = [1/(b-a)] \int_a^b \text{diam } \gamma[I(z, t)] \, dz$$

$$= (1/Nt) \int_a^b \text{diam } \gamma[I(z, t)] \, dz \tag{2.41}$$

and therefore, with $\xi(z, t) = \text{diam } \gamma[I(z, t)]$, the length $l(t)$ of Γ is

$$l(t) = N \cdot \xi = (1/t) \int_a^b \xi(z, t) \, dz \tag{2.42}$$

It is important to note the difference between this method and that based on approximation by arcs of equal length. Here, $\xi(z, t)$ varies along Γ; if the projectile passes the point $z_\Gamma = \gamma(z)$ at high speed, $\xi(z, t)$ is large and the curve is regular, whilst if its speed is low then ξ is small and the curve is locally chaotic, much folded up on itself as in Figure 2.1. Thus $\xi(z, t)$ is a *measure of local irregularity*: but we must not forget that all this depends on the mapping γ.

2.3.2 *Length of a rectifiable curve*

By definition, a curve is rectifiable if its length $l(t)$ tends to a finite limit as $t \to 0$, and in this case it is also derivable. The standard method for measuring the length of such a curve is to step along it with dividers, which corresponds physically with *traversing the trajectory at constant speed*. The Hausdorff measure has the same significance: a very contorted part of the curve will give rise to a covering disc of small diameter, whilst a smooth part will have a disc of large diameter: again, see Figure 2.1.

As an example, consider the graph of a continuous derivable function $y = f(z)$, over an interval $[a, b]$ of z. The natural parametrisation is by z, and with this $\xi(z, t)$ is the diameter of that part of the graph for which the abscissae are between $z - t/2$ and $x + t/2$. For t sufficiently small this arc can be taken as a linear segment with end-points $[z \pm t/2, f(z \pm t/2)]$ and therefore of length

$$\xi(z, t) = \sqrt{[t^2 + \{t \cdot f'(z)\}^2]} = t\sqrt{[1 + \{f'(z)\}^2]} \tag{2.43}$$

Therefore

$$l(t) = \lim_{t \to 0} (1/t) \int_a^b \xi(z, t)\, dz = \int_a^b \sqrt{[1 + \{f'(z)\}^2]}\, dz \tag{2.44}$$

which is the standard formula giving the length of a rectifiable curve in terms of its derivative.

2.3.3 *The case of non-rectifiable curves*

In the above we have shown that provided that Γ is derivable the ratio $t/\xi(z, t)$ has a finite limit as t tends to zero; in fact in the limit $\xi(z, t)$ is identical with the differential element of Γ. This is no longer the case when the curve is not rectifiable, and our main interest in this study is in non-rectifiable curves which are self-similar fractals, that is, curves for which the logarithmic density tends to a constant value:

$$\log t / \log \xi(z, t) \to \Delta \quad \text{as} \quad t \to 0 \tag{2.45}$$

The limit Δ is the fractal dimension of the curve, and the length $l(t)$ tends to infinity, for from the previous analysis it follows that

$$l(t) \sim t^{(1/\Delta) - 1} \tag{2.46}$$

An example is the von Koch curve, for here the limit is $\log 4 / \log 3$, approached uniformly in the interval $[0, 1]$.

It would be interesting to generalise the result expressed by equation 2.34 by showing under what conditions

$$\Delta = 1/[1 + \lim_{t \to 0} \{\log l(t)/\log t\}] \tag{2.47}$$

(a relation in which the length appears explicitly), noting that in the case of an internal similitude, $\xi(z, t) \sim t^{1/\Delta}$, i.e. $\log t \sim \Delta \log \xi$, this becomes

$$\Delta = 1/[1 + (1/(\Delta)) \lim_{\xi \to 0} \{\log l(t)/\log \xi\}] \tag{2.48}$$

whence

$$\Delta = 1 - \lim_{\xi \to 0} [\log l(t)/\log \xi]$$

a familiar formula of Chapter 1 (cf. Equation 1.2).

Whilst this last expression gives a relation between $l(t)$ and ξ this is not unique, for it depends on the space-time coupling and the mapping γ. We shall give an example, but first we consider the application of the Fourier transform to this space-time coupling.

An aspect of particular importance for applications in physics is the following. Applying the Laplace transform to $l(t) \sim t^{(1/\Delta - 1)}$ we obtain for the space-measuring gauge

$$\eta(p) \sim (1/p)^{1/\Delta} \tag{2.49}$$

i.e.

$$[\eta(p)]^\Delta \sim 1/p \tag{2.50}$$

where p, the transform parameter, has here the same meaning as the frequency variable in the sequence used in § 1.1 to define the polygonal approximation to the curve.

Returning to t-space this is equivalent to $\xi \sim t^{1/\Delta}$. These relations show that there is an unavoidable relation between time and space which means that space must not be used as a gauge without first ensuring that the time associated with this can also be used as a gauge. This relation, which is conveyed by means of the fractal dimension of the space, is simply the ergodic relation in a fractal

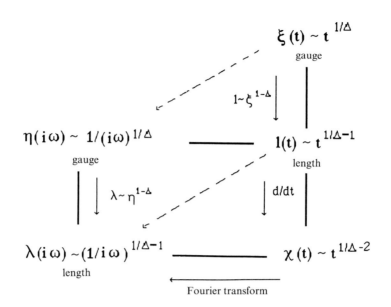

Figure 2.6 Relations between characteristic lengths and time scales on the one hand and frequencies on the other, for the measurement of a self-similar curve in Fourier space and in time space (Laplace transform space is used in the text).

environment; it expresses an *uncertainty* similar to that of quantum mechanics, or a *dispersion* as in wave theory.

The important point here, which is not at all obvious, is that the Fourier and Laplace transforms transform one scale law into another. As Figure 2.8 shows, there is an isomorphism, so far as scale laws are concerned, between the direct space and the reciprocal (Fourier or Laplace) space. This property simplifies considerably the analysis of fractal objects, and makes possible the development of the TEISI model. It undoubtedly underlies the universality of fractal geometry in nature.

2.3.4 *Fractal parametrisation of a self-affine graph*

The analysis of the preceding paragraph has concerned the parametrisation of a self-similar curve, and the results obtained are valid only for this case; so the question arises, what corresponding results, if any, are there for a self-affine graph? The answer is that if Γ does not have internal similarity *the most natural parameter is the abscissa* (cf. Fig. 2.7). Using the variation method applied to

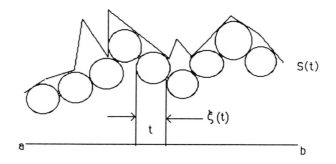

Figure 2.7 Measurement of a self-affine graph. Note that the method leads to an estimate for the neighbourhood of the graph and hence to the co-dimension.

graphs we shall establish this, and show that this parametrisation is distinct from the preceding form.

Let f be a continuous function defined on [a, b]; as in para. 2.2.2, we define the *t-oscillation* at z as the maximum variation of f(z) in the interval dz (cf. Fig. 2.4) and the *t-variation* as the integral of this over [a, b]:

$$V_t = \int_a^b v_t \, dz \tag{2.51}$$

The self-affine property of the graph now allows us to take $\xi(z, t) \sim t$, a very different choice from the previous $\xi \sim t^{1/\Delta}$. We have therefore

$$v_{t/2} < \xi(z, t) < v_{t/2} + t \tag{2.52}$$

giving

$$(1/t) V_{t/2} < l(t) < (1/t) V_{t/2} + (b - a) \tag{2.53}$$

The rate of growth of l(t) is thus the same as that of $(1/t) V_\Delta$, so from the result obtained previously

$$\Delta_{MB} = \lim_{t \to 0} [2 - \log V_t / \log t] \tag{2.54}$$

we have, for a self-affine graph,

$$\Delta_{MB} = \lim_{t \to 0} [1 - \log l(t) / \log t] \tag{2.55}$$

i.e.

$$l(t) \sim t^{1-\Delta_{MB}} \tag{2.56}$$

This is a very different relation from that obtained for sets with internal similarity. The two give results that are approximately the same for a fractal of dimension Δ_Γ not very different from 1, in a 2-dimensional space, but are very different for $\Delta_{MB} > 1$. Clearly, the reason for this must lie in the different parametrisations of the curves. Thus in fractal geometry the type of coupling between space and time, even in the limit, is not fixed but depends on the mapping γ.

In physical terms this mapping corresponds to what is traditionally called the measuring process. The above analysis shows that there are two large classes of measure in terms of which the space gauge and the time gauge are coupled linearly and via a fractal geometry respectively. The first corresponds to measure by an observer in Euclidean space, a fractal interface occurring only at the boundary, as in the TEISI model; the second to the embedding of the physical process, time included, in the fractal geometry. Confusion between these types of measure, which we may call external and internal respectively, is responsible for many of the difficulties in understanding the physics of fractal media.

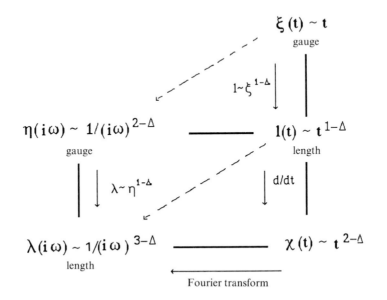

Figure 2.8 Relations as in Fig. 2.6, here for a self-affine curve.

2.4 Case where the geometry is a transfer function: physical measure of the geometry

The two results just obtained may seem strange and contrary to commonsense:

(a) They ascribe a time-dependence to a length; that is, they imply an essential coupling between space and time, with time being understood as a parameter.

(b) They have no relation to any specific physical phenomenon, since the mapping γ is arbitrary. In particular, they have no concern with diffusion: the relation $\langle \xi^2 \rangle \sim t$, the usual starting point for the analysis of any dissipative physical process controlled by transport, appears here simply as a special case, the Peano curve.

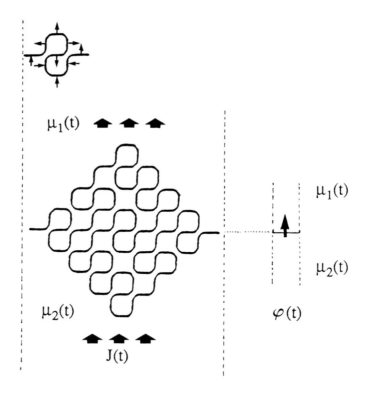

Figure 2.9 Linear relation between a "flux" and a "flux density", linked by a "length" which is a function of the generalised frequency. In the TEISI model this represents a diffusion process (cf. para. 7.8).

Returning to the expression for l(t) for a self-similar geometry, this (Equation 2.34) can be written

$$l(t) \sim [t^{1/\Delta}]^{1-\Delta} \sim [\xi]^{1-\Delta} \tag{2.57}$$

We shall show in para. 5.1 that every measure for the support for a distribution requires the use of a test function $\varphi(t)$ associated with a physical process; and the same holds for the measure of l(t). The efficiency of physical processes is seldom 1 and the origin of irreversibility lies in the macroscopic exchange fluxes J(t) across boundaries, the fractal geometry of which now shows us that they can be fluid. This leads to the question: could irreversibility be given a geometrical content? and if so, what would be the role of time, considered as a parameter?

To investigate this we start with the classical relation between flux J(t) and flux density $\varphi(t)$, $J(t) = L\varphi(t)$, where L is the length of the interface over which the exchange takes place. Whilst it may seem surprising, a simple way to adapt this to the case where L is a function l(t) of time is to give the geometry the property of a transfer function, and thus to form the convolution of $\varphi(t)$ with dl(t)/dt:

$$J(t) = dl(t)/dt * \varphi(t) = \chi(t) * \varphi(t) \tag{2.58}$$

(cf. Fig. 2.9 and § 5.1; * is the convolution operator).

With respect to the measure of the fractal set Γ, $\varphi(t)$ plays the role of test function (see 5.1), and in the p-space of the Laplace transform

$$J(p) = \lambda(p) \cdot \varphi(p)$$

A reader who finds this surprising should study the first page of the chapter on Applications, where he will find that the theory of distributions brings together the above operations under the heading of *regularisation of a fractal interface by Dirac approximations*. If we assume self-similarity for the interface we have $\chi(t) \sim t^{(1/\Delta)-2}$ and therefore

$$J(t) \sim t^{(1/\Delta)-2} * \varphi(t)$$

the Laplace transform of which gives

$$p^{(1/\Delta)-1} J(p) \sim \varphi(p)$$

From this, we can write the relation between J(t) and $\varphi(t)$ as

$$\mathscr{D}_t^{(1/\Delta)-1} J(t) \sim \varphi(t) \tag{2.59}$$

where the notation $\mathscr{D}^\alpha$ denotes a derivative of non-integral order α, a concept developed in Chapter 3. If we now take for the measure the derivative of the flux density, $\mu(t) = \mathscr{D}\varphi(t)$, we have for the state function $C(p) \sim J(p)/\mu(p)$

$$C(p) \sim 1/p^{1/\Delta} \tag{2.60}$$

This state function is also the transfer function (impedance) for an object which we may call a *fractance*, because on the one hand it is associated with the fractal character of a self-similar interface and, on the other, it provides a formal generalisation of the electrical property of capacitance.

The reasoning we have just given is extremely general, and owes its value to the mathematical clarity with which the concept of parametrisation has been introduced. The method used underlines the basic role of the mapping γ, and also to some extent the arbitrariness of this, resulting from the freedom of choice. The results are fundamental for understanding dissipation in a fractal environment and its dynamics in terms of fractal geometry – the TEISI model and δ-transfer.

The analysis so far has been for a self-similar curve; it can be extended to the case of a self-affine graph in the plane as follows.

We now have $\chi(t) \sim t^{1-\Delta}$, so from 2.47 $J(t) \sim t^{-\Delta} * \varphi(t)$, from which, by the same process as before, we get

$$\mathscr{D}_t^{1-\Delta} J(t) \sim \varphi(t) \tag{2.61}$$

and finally

$$C(p) \sim 1/p^{2-\Delta} = 1/p^{\zeta} \tag{2.62}$$

This state function is called the *co-fractance*; the necessary conditions for the validity of the relation are given in para. 7.1.

The results obtained here differ according as the curve is self-similar or self-affine. The reason lies in the different natural measures for the two types: for the self-similar curve this is by arc lengths, whilst the self-affine curve requires a method of approximation in which the co-dimension ζ plays a leading role.

If in $C(p)$ we put $p = i\omega$ we see that $C(p) = (i\omega)^{-\alpha}$ has an associated phase angle of absolute value $(1-\alpha)\pi/2$, independent of the frequency ω. This phase shift, controlled by the fractal dimension via the parametrisation, is in fact at the heart of the irreversibility imposed by the test function $\varphi(t)$.

This measure of length, which is both transfer function and state function, leads naturally to the use of derivatives of non-integral order. There is thus a relation between the parametrisation of a fractal curve, derivatives of non-integral order and the irreversibility of physical processes involving exchanges over a fractal interface. This is taken up in more detail in Chapter 3.

Note: the idea of making geometry the support for the concept of dissipation, and thus of irreversibility, is not new. The impedance of a linear transformation represented by a matrix can be defined as the ratio of the determinant of the matrix to that of its first minor. This concept has been used widely in electricity and can be related directly to that of volume and surface, or, more generally, to differentiable varieties of dimension d and d − 1.

Chapter 3

Derivatives of non-integral order

The analysis in the previous chapter has led to a clear distinction between the concept of a fractal metric, of which the fractal dimension is the characteristic feature, and parametrisation; and we have seen that the latter can be of internal (based on a state variable) or external (based on an environmental variable and expressed by the co-dimension) origin or can be arbitrary. At the same time, experimental studies have shown a need to appeal to a concept not yet very familiar in physics, that of a *derivative of non-integral order*, which we shall call *non-integral (or fractional) derivation*. It seems that this operation is, in particular, the key to the uncertainty relation in physics and, through the flux/forces relations, to the understanding of irreversibility. We denote the relevant operator by $\mathscr{D}_t^{\alpha} = \mathrm{d}_{\alpha}/\mathrm{d}t_{\alpha}$; the order α can be a complex number, but here we shall confine ourselves to real values.

The physical meaning of a derivative of order 1/2 is already known: as Oldham and Spanier have shown, it is the operator representing diffusion in a semi-infinite medium (see para 7.8). The general case (for real order) was the theme of a special conference held in Newhaven in 1974, motivated by the feeling that since a meaning was readily attached to derivation of order 1 (the slope of a curve) and of -1 (an integral), it should be possible to assign a meaning to a general non-integral order; but the conference ended with the participants declaring that this was not possible. The physicists who took part were not aware of the geometrical representation underlying the work of Gemant, who in 1936, for the first time so far as we know, and without any reaction from the scientific world, showed the relevance of fractional order derivation to the physics of non-linear phenomena: the phenomenon he studied was the visco-elasticity of bread dough!

Renewed interest in the geometry of non-differentiable curves, stimulated by Mandelbrot's work, has in recent years led to the same question being raised from a slightly different point of view, that of the existence of the derivative. As we have seen, if the derivative of order 1 exists at a certain point, it is locally rectifiable there, and as a first approximation can be treated as a straight-line segment. For a continuous but nowhere derivable function, however, the graph is extremely irregular and the length in the neighbourhood of almost every point is infinite: there is thus some relation between non-derivability and non-rectifiability. Further, fractal geometry, through the concept of dimension to which it attaches such importance, provides a means for quantifying complexity; so it should be possible to relate dimension and order of derivation: the question is, how?

In 1982, eight years after the Newhaven conference, one of the first international conferences on fractal geometry was held in Courchevel, at which the present author, reporting the results of five years' research, was able to display fractal objects of truly engineering origin – battery electrodes, in fact – on which actual experiments could be performed. He was able to state that in an ideal case certain physical properties associated with irreversibility, energy flow for example, are directly related to non-integral derivation of the support for the dissipation. He showed that events within the electrode occurred as though in a time dimension that differed from that of the observer.

Experimental work has shown that the link between fractality and non-integral derivation is essential to a proper approach to reality in many fields where fractality plays a role and introduces scaling laws. The present chapter, based on the work of C. Tricot, aims to establish this theoretically, starting from particular cases, and the analysis is supported by graphs of continuous non-derivable functions. We shall show that non-integral derivation provides a very efficient method for measuring the degree of irregularity of a fractal curve and for relating this to the noise that affects the measurement of any phenomenon that originates in this irregularity.

3.1 Definition of derivation of non-integral order

The idea of interpolating the infinitely small goes back to Euler (1760), but the methods used to-day were established only last century, by Liouville and Riemann. The idea of this interpolation in practice originated in the study of the derivative of the power of a variable. It should cause no surprise to see the concept of the non-integral derivative appearing in connection with the scaling laws that are expressed by such functions. As an example, we start with the derivative of t^p.

3.1.1 *Derivative of* t^p

Let $f(t) = t^p$, $p > -1$; it is well known that for all positive integers n

$$\mathscr{D}_t^n f(t) = p(p-1)\ldots(p-n+1)t^{p-n}$$

The derivative of order 1 represents a "speed" with respect to the variable t, of order 2 an "acceleration" or a "speed of the speed", and so on.

The converse operation, integration, is expressed by

$$\int_0^t f(v)\,dv = t^{p+1}/(p+1)$$

and n successive integrations give

$$\int_0^t \int \ldots \int f(v)\,dv = t^{p+n}/(p+1)(p+2)\ldots(p+n)$$

We can interpolate between these two classical operations by using the Euler gamma-function, which is a generalisation of the familiar factorial function, defined only for positive integer arguments, to any real argument.

3.1.2 *The Euler gamma function* (Γ)

This is defined by

$$\Gamma(t) = \int_0^\infty e^{-u}u^{t-1}\,du, \quad t > 0 \tag{3.1}$$

It is well known, and is easily shown by integration by parts, that the function satisfies the fundamental relation $\Gamma(t+1) = t\Gamma(t)$. Since $\Gamma(1) = 1$, it follows that for any positive integer n, $\Gamma(n+1) = n!$.

The equation 3.1 defines $\Gamma(t)$ only for $t > 0$, and gives $\Gamma(t) \to \infty$ as $t \to 0$; the definition is extended to all real values of t by means of the relation $\Gamma(t+1) = t\Gamma(t)$, from which we get

$$\Gamma(t) = \Gamma(t+n)/t(t+1)\ldots(t+n-1), \quad \text{for} \quad t+n > 0, \quad \text{i.e.} \quad n > -t$$

This defines $\Gamma(t)$ for all real t except zero and negative integers, for which values $\Gamma(t) = \to \pm\infty$.

Armed with this, we now *define* the derivative of non-integral order of t^p as follows:

$$\mathscr{D}_t^\alpha t^p = [\Gamma(p+1)/\Gamma(p+1-\alpha)]t^{p-\alpha}, \quad p > -1$$

The integral can be regarded as the derivative of order $\alpha = -1$.

The next case to consider is the derivation of the exponential function, which is of such frequent occurrence in physics.

3.1.3 *Derivative of e^{pt}*

For all positive integers n

$$\mathscr{D}_t^n e^{pt} = p^n e^{pt}$$

and since

$$\int_{-\infty}^t e^{pu} \, du = p^{-1} e^{pt}$$

the result of n successive integrations is $p^{-n} e^{pt}$.
Thus it is natural to define

$$\mathscr{D}_t^\alpha e^{pt} = p^\alpha e^{pt} \quad \text{for all real } \alpha$$

Then if $f(t)$ can be developed as a uniformly convergent series of exponentials

$$f(t) = \sum_{n=0}^\infty c_n e^{npt}$$

$$\mathscr{D}_t^\alpha f(t) = \sum_{n=0}^\infty c_n (np)^\alpha e^{npt} \tag{3.2}$$

If p is a pure imaginary, $p = i\omega$ say, then $e^{pt} = \cos\omega t + i\sin\omega t$, so taking real and imaginary parts

$$\mathscr{D}_t^\alpha \cos/\sin \omega t = \omega^\alpha \cos/\sin(\omega t + \alpha\pi/2) \tag{3.3}$$

This reproduces the familiar results when α is a positive integer, e.g. for $\alpha = 1, (d/dt)\cos pt = -p\sin pt$; and it will be noticed that the operation results not only in a change of amplitude (the factor p^α) but also a change of phase (a shift of $\alpha\pi/2$), the latter being independent of the frequency ω.

However, this definition leads to an inconsistency, as we now show.

3.1.4 *An inconsistency*

The function e^{pt} has a power series development $\sum\limits_{n=0}^{\infty} (pt)^n/n!$ and if we operate on this term by term, putting $n! = \Gamma(n+1)$,

$$\mathscr{D}_t^\alpha e^{pt} = \sum_{n=0}^{\infty} p^n t^{n-\alpha}/\Gamma(n+1-\alpha)$$

which clearly differs from the above result

$$\mathscr{D}_t^\alpha e^{pt} = p^\alpha e^{pt} = \sum_{n=0}^{\infty} p^{n+\alpha}t^n/\Gamma(n+1)$$

What might seem here to be a major contradiction is easily resolved by noting that there are at least two definitions of the fractional order derivative – in fact, an unlimited number can be constructed – depending on the value of a certain parameter; this has not been taken into account in the above reasoning, because the non-local character has not been treated rigorously. Thus, following the various approaches made in the 19th century, we are interested only in the dependence as a function of time, without specification precisely in each case the lower bound of the integral. We shall see later that the value taken for this affects the resulting expression. Further, the validity of term-by-term derivation has not been established: it is well known that whilst a series may be convergent the series obtained by term-by-term differentiation (in the ordinary sense of the word) need not converge. In the context we are considering it is only the lower bound of the integral that interests the physicist, since it is this that limits the memory of the inheritance equations, governed by differential equations of fractional order.

3.1.5 *Generalisation: the Riemann-Liouville integral*

Dirichlet showed that the n'th integral of a function $f(x)$ between the limits a and t can be written as a convolution:

$$\mathscr{D}_x^{-n}f(x) = [1/\Gamma(n)] \int_a^t f(y)(t-y)^{n-1}\,dy = [1/\Gamma(n)]f(t) * t^{n-1}$$

This suggests a generalisation of the integration operation to one of non-integral order, that is, of the derivation operation to arbitrary negative order $-\alpha(\alpha > 0)$. Writing this as $_a\mathscr{D}_t^{-\alpha}$ we then have

$$_a\mathscr{D}_t^{-\alpha}f(t) = [1/\Gamma(\alpha)] \int_a^t f(y)(t-y)^{\alpha-1}\,dy \tag{3.4}$$

The integral on the right is known as the Riemann-Liouville integral.

In this definition we have not one but two parameters, the order α and the lower bound a of the integration. To clarify their respective roles, consider again the particular cases of t^p and e^{pt}.

With a = 0, $f(t) = t^p$ and $u = y/t$ the convolution integral is

$$_0\mathscr{D}_t^{-\alpha}(t^p) = [1/\Gamma(\alpha)]\,t^{p+\alpha} \int_0^1 u^{p+1}(1-u)^{\alpha-1}\,du$$

and since $\int_0^1 u^{p+1}(1-u)^{q-1}\,du = B(p,q) = \Gamma(p)\Gamma(q)/\Gamma(p+q)$ (the beta function) this gives

$$_0\mathscr{D}_t^{-\alpha}(t^p) = [\Gamma(p+1)/\Gamma(p+1+\alpha)]t^{p+\alpha}$$

Thus for t^p at least, this definition of integration of non-integral order gives the result one would expect, provided that the lower limit of the integral is 0.

It will be seen that this expresses the operator $_0\mathscr{D}_t^{x-\alpha}$ in terms of a *convolution with a power-type function* $X_{\alpha-1}(t) \sim t^{\alpha-1}$, thus giving a scaling law. In the theory of distributions this convolution is a *smoothing* (cf. 5.1.1):

$$f(t) * X(t) = \int_0^t f(y)X(t-y)\,dy \tag{3.5}$$

If we now consider the integration of e^{pt} and take for the lower limit of the integral not 0 but $-\infty$, we have

$$_{-\infty}\mathscr{D}_t^{-\alpha}(e^{pt}) = [1/\Gamma(\alpha)] \int_{-\infty}^{t} e^{py}(t-y)^{\alpha-1}\,dy \tag{3.6}$$

which was shown by Krug (1890) to be $p^{-\alpha}e^{-pt}$.

If $X_\alpha(t) = t^\alpha$, then for any function $f(t)$

$$_{-\infty}\mathscr{D}_t^{-\alpha}f(t) = [1/\Gamma(\alpha)]\,[X_{\alpha-1}(t) * f(t)] \tag{3.7}$$

Weyl made a particular study of the operator $_{-\infty}\mathscr{D}_t^{-\alpha}$, which has consequently been called the *Weyl integral*.

3.1.6 *Definition of the integro-differential operator*

The above analysis has given definitions of integration of non-integral order for the functions t^p and e^{pt} which are mutually consistent, each corresponding to a Riemann-Liouville integral but with different lower limits, a and $-\infty$ respectively. The following is the notation most commonly used.

For interpolating the integral:

$$_{a}\mathscr{D}_t^{-\alpha}[f(t)] = [1/\Gamma(\alpha)] \int_{a}^{t} f(y)(t-y)^{\alpha-1}\,dy, \quad \text{where} \quad a > 0$$

For interpolating the derivative we need only to differentiate this relation with respect to t: so if $0 < \alpha < 1$

$$_{a}\mathscr{D}_t^{\alpha}[f(t)] = [1/\Gamma(1-\alpha)]\mathscr{D}_t^{1}\left[\int_{a}^{t} f(y)(t-y)^{-\alpha}\,dy\right] \tag{3.8}$$

and if $n-1 < \alpha < n$, we can replace α by $\alpha-(n-1)$, giving

$$= [1/\Gamma(n-\alpha)]\mathscr{D}_t^{n}\left[\int_{a}^{t} f(y)(t-y)^{n-\alpha-1}\,dy\right] \tag{3.9}$$

since then $0 < \alpha-(n-1) < 1$.

If the derivative $_{a}\mathscr{D}_t^{\alpha}[f(t)]$, $\alpha \in \mathbb{R}$, is defined for t then we say that f is derivable with order α at t.

This analysis has been undertaken to ensure that the operators are continuous functions of a, that they can be permuted if the order α is negative and they have the same lower limit, that there are formulae corresponding to that of Leibnitz, and so on. Given this, we can use, without the need for further proof, the most useful properties of derivation of non-integral order. We shall not pursue this further here, but refer the reader to the Bibliography for the literature of the subject, which in many cases gives excellent historical and biographical surveys.

The following are a few properties which it is useful to keep in mind.

(1) $_a\mathscr{D}_t^0[f(t)] = f(t)$

(2) If n is a positive integer, $_a\mathscr{D}_t^n[f(t)] = \mathscr{D}_t^n[f(t)]$, the n'th derivative in the usual sense, independent of a; and $\mathscr{D}_t^{-n}[f(t)]$ is the n'th repeated integral, provided that all the intermediate integrals vanish at $t = a$.

(3) $_a\mathscr{D}_t^\alpha[p \cdot f(t) + q \cdot g(t)] = p \cdot {_a\mathscr{D}_t^\alpha}[f(t)] + q \cdot {_a\mathscr{D}_t^\alpha}[g(t)]$ for all p, q   (3.10)

(4) $_a\mathscr{D}_t^\alpha \cdot {_a\mathscr{D}_t^\beta}[f(t)] = {_a\mathscr{D}_t^\beta} \cdot {_a\mathscr{D}_t^\alpha}[f(t)]$ (3.11)

These operators can be used to obtain elegant and compact expressions for certain differential equations and/or their solutions, such as those that arise in diffusion theory and in problems that lead to special functions of the type of Bessel functions. Their present use seems to be always for purely analytical reasons and so far as we know nothing has been done to attach any physical meaning. We now give a short treatment of Bessel functions, for which, it seems to us, some effort should be made in that direction.

3.1.7 *Bessel functions*

These functions, which play an important role in mathematical physics, are solutions of the differential equation

$$f'' + (1/x)f' + (1 - v^2/x^2)f = 0$$

where f'', f' are respectively the first and second derivatives of f with respect to x and v is an arbitrary complex number.

Imposing the condition that the solution is finite (including zero) at $x = 0$ leads to the *Bessel functions of the first kind of order v*, denoted by $J_v(x)$, which can be expressed in various ways: e.g. as a power series

$$J_\nu(x) = \sum_{k=0}^{\infty} (-1)^k (x/2)^{2k+\nu}/(k!)\Gamma(k+\nu+1)$$

or as an integral

$$J_\nu(x) = [x^{-\nu} 2^{1-\nu}/\sqrt{\pi} \cdot \Gamma(\nu+1/2)] \int_0^x (x^2 - t^2)^{\nu-1/2} \cos t \, dt$$

A less frequently used representation is by what is called Rodrigues formula, which uses derivation of non-integral order;

$$J_n(x) = [(2x)^{-n}/\sqrt{\pi}] \mathcal{D}_u^{-n-1/2}[(\cos x)/x]$$

where $u = x^2$ and $\nu = -n - 1/2$. This is not valid for n integral, but with an appropriate definition of fractional-order derivation the function can be written

$$J_{-\nu-1/2}(x) = (2x)^{\nu+1/2}\pi^{-1/2}\mathcal{D}_u^\nu[(\cos x)/x]$$

where again $u = x^2$.

Such a representation can be very useful, because it can suggest manipulations that would not be suggested by other forms; for example, from the general relation

$$\mathcal{D}^\alpha \cdot \mathcal{D}^\beta = \mathcal{D}^{\alpha+\beta}$$

we get

$$\int_0^x J_{-\beta-1/2}(t) \cdot (x^2 - t^2)^{-\alpha-1} \cdot t^{-\beta+1/2} \, dt = 2^{-\alpha+1/2} x^{-\alpha-\beta-1/2} J_{-\alpha-\beta-1/2}(x)$$

a result which would not be easily obtained by the standard methods.

3.2 Spectral analysis and non-integral derivation

We now show that if f(t) is defined over the positive reals there is a relation between its Fourier spectrum and its fractional derivation. There is in fact a relation between fractional derivation, convolution by the geometry via the parametrisation, and the phase change. The Fourier transform being a smoothing, we can expect there to be a relation between this and fractional derivation.

3.2.1 *Fractional derivation and the spectrum*

We have shown (equation 3.7) that if $0 < \alpha < 1$, then $_a\mathscr{D}_t^{-\alpha}f(t)$ is proportional to the convolution $X_{\alpha-1} * f$; so if we write T_F to denote the Fourier transform,

$$T_F[_a\mathscr{D}_t^{-\alpha}f(t)] \sim T_F[X_{\alpha-1}(t)*f(t)] \sim T_F[X_{\alpha-1}] \cdot T_F[(f)]$$

and since $X_{\alpha-1} \sim t^{\alpha-1}$, $T_F[X_{\alpha-1}] \sim (i\omega)^{-\alpha}$; thus

$$T_F[_a\mathscr{D}_t^{-\alpha}f(t)] \sim (i\omega)^{-\alpha}T_F[f(t)]$$

It follows that if ω is the characteristic frequency associated with the variable t

$$T_F[_a\mathscr{D}_t^{\alpha-1}f(t)] \sim (i\omega)^{\alpha-1}T_F[f(t)] \qquad (3.12)$$

Thus, apart from a numerical coefficient, the non-integral derivation of the graph of a function multiplies its Fourier transform by $(i\omega)^{-\beta}$. If the energy spectrum $E[f(t)] = \{T_F[f(t)]\}^2$ exists, that of its fractional derivative is proportional to $\omega^{-2\beta}E[f(x)]$ – that is, the energy spectrum is divided by $\omega^{2\beta}$. It is precisely this property that makes it possible to construct the Werner-Levy process, having a spectrum $1/\omega^2$, from *white noise*, with spectrum $1/\omega^0$; and more generally, fractional Brownian-motion noise with spectrum $1/\omega^{2H+1}$, to use the standard notation.

This elementary argument shows that integro-differential operations of non-integral order can be expressed very simply in Fourier space. It should however be noted that the quantum of time, normally defined by the frequency, has somehow been "broken", since everything is happening as though there were a fractional frequency.

We next give an example to illustrate the relation between fractional-order derivation and the regularity of a curve, based on work on rough surfaces by C. Tricot and others in collaboration with the tribology group at ESMM, Besançon.

3.3 An irregularity parameter for continuous non-derivable functions: the maximum order of derivation

The general works that treat fractional derivation deal mainly with the algebra over the class of these operators, the extension of the classical theory of differentials, the calculation of particular derivatives and the possible use of the techniques in attacking problems of analysis. They give little or no consideration to what is our first concern here, the question of the actual existence of these derivatives. We shall show that the derivative of non-integral order can exist for a function that is not derivable to integral order, provided that the non-integral order is in a certain relation to the fractal dimension of the graph of the function. Our problem is then to find that relation.

3.3.1 *Existence of the derivatives*

There are two questions:

1. If the derivative of order α, $_a\mathscr{D}_t^\alpha f(t)$, exists for a given non-integral α, does the same derivative of order β exist for all $\beta < \alpha$?

2. If as before $_a\mathscr{D}_t^\alpha f(t)$ exists, does $_b\mathscr{D}_t^\alpha f(t)$ exist for all b?

Before attempting to answer these, consider the following examples.

A trigonometric series

Suppose f(t) can be expressed in the form

$$f(t) = \sum_\omega c_\omega e^{i\omega t} \tag{3.13}$$

where t is a real variable and the c_ω can be complex: thus f can take complex values. Formally, the Weyl integral (cf. § 3.1.5) of f can be written

$$_{-\infty}\mathscr{D}_t^\alpha f(t) \sim \sum_\omega c_\omega (i\omega)^\alpha e^{i\omega t}$$

It can be shown, by using Abel's theorem, that this converges, and therefore for all $\beta < \alpha$

$$_{-\infty}\mathscr{D}_t^\beta f(t) \sim \sum_\omega c_\omega (i\omega)^\alpha (i\omega)^{\beta-\alpha} e^{i\omega t}$$

also converges; and therefore if the derivative of order α exists, so does that of any order $\beta < \alpha$.

Order of derivation	$\beta < \alpha$	α	$\beta > \alpha$
Existence	YES	YES	UNKNOWN

Table 3.1

However, this argument tells us nothing about the case $\beta > \alpha$, and it is possible that there is a *maximum order of non-integral derivation*. If there is, its value can serve as a criterion for the "regularity" of a curve that is non-differentiable in the usual sense: of two curves, the one with the higher maximum order of non-integral derivation could be said to be the more "regular".

A function whose integral is bounded

A partial answer to the second question is:

   The existence of the Weyl integral is a necessary condition for the existence of all other derivatives with finite lower limits: i.e.

   $_a\mathscr{D}_t^{\alpha}f(t)$ exists for all $t > a$ if $_{-\infty}\mathscr{D}_t^{\alpha}f(t)$ exists.

   Thus the existence of the Weyl integral, by virtue of the value of its lower limit, justifies non-integral derivation whatever the value of the lower limit. Since the function $e^{i\omega t}$ fulfils the condition for any real ω, we can say that the derivatives of cos/sin ωt, of any real order α, exist, whatever value is chosen for the lower limit of the integral.

   The reader may find these questions very formal, perhaps even unnecessary; but they are relevant to the possible practical uses of fractal geometry. In fact, just as we saw in Chapter 2 (§ 2.1.1) that a fractal dimension can be understood as a cut dividing the real numbers into two classes, Table 3.1 shows the possibility of regarding the upper bound to the order of derivation also as a cut in the reals; and it is natural to raise the question of the existence and meaning of any relation there might be between these two cuts. But first we note a sufficient condition for the validity of derivation of all orders.

$$\text{Let} \quad X_p(t) = \begin{cases} 0 & \text{if } t \leqslant 0 \\ t^p & \text{if } t > 0 \end{cases}$$

where p is in the open interval $]0, 1[$. All fractional derivatives of X_p exist, and since the function vanishes on the left of the origin

$$_{-\infty}\mathscr{D}_t^\alpha X_p(t) = {}_0\mathscr{D}_t^\alpha X_p(t) = [\Gamma(p+1)/\Gamma(p+1-\alpha)]t^{p-\alpha}$$

X_p does not have a derivative of order 1 at $t = 0$, so the question arises, given the value of p, for what value(s) of α does the derivative exist there? The answer is that *it exists for all $\alpha < p$*. For example, $X_{1/2}(t) = t^{1/2}$ has fractional derivatives of order up to but not including $1/2$.

Whatever the limit thus set to the value of α, nothing prevents us from taking derivatives of any integer order: thus $\mathscr{D}_t^1 t^p = p \cdot t^{p-1}$. In other words, whilst it is not valid to take fractional derivatives of t^p of order p or greater, derivation to greater integral orders is allowed. Thus, finally, we can say that *this cut separates the class of derivatives of continuous orders from those of discrete orders.*

3.2.2 *Extremum for the order of derivation, and the fractal dimension*

Let $\delta(f, t)$ denote the upper bound of the set of non-integral values of α:

$$\delta(f, t) = \sup[\alpha] \quad \text{such that} \quad {}_a\mathscr{D}_t^\alpha f(t) \text{ exists}$$

At least for functions such that the answer to Question 2 of §3.3.1 is affirmative, $\delta(f, t)$ is independent of the lower limit a of the integral. Since we are especially interested in functions which have some form of homogeneity, for example with scaling laws, we consider $\delta(f, t)$ as a function of t; it will have a lower bound for t, which we write

$$\delta(f) = \inf \delta(f, t) \tag{3.14}$$

The interpretation of this is that for all values of t, the derivative of non-integral order α exists for all $\alpha < \delta(f)$:

$$\text{for all t,} \quad {}_a\mathscr{D}_t^\alpha f(t) \quad \text{exists for all} \quad \alpha < \delta$$

The next question to ask is, what relation, if any, is there between δ and the fractal dimension of the curve $[f(t), g(t)]$ in the case of a self-similar curve, or $f(t)$ in the case of a self-affine graph? As an example, consider the case of a continuous non-derivable function, the Weierstrass function $W(t)$ of §5.1.2:

$$W(t) = \sum_{n=0}^{\infty} \omega^{-nH} \cos \omega^n t \tag{3.15}$$

W(t) belongs to the class of functions described as "noise of type $1/f$"; here $f = \omega^{2H}$, the factor 2 in the exponent relating to the energy spectrum; the amplitude is characterised by $1/\omega^H$.

We investigate the critical value of α, defining the transition from non-derivability (of fractional order) to derivability; generalising to complex values of W(t):

$$_{-\infty}\mathscr{D}_t^{\alpha-1}W(t) = \sum_{n=0}^{\infty} \omega^{-nH}(i\omega^n)^{\alpha-1}\exp(i\omega^n t)$$

$$= \sum_0^{\infty} \omega^{-n(H-\alpha+1)}\exp[i(\omega^n t+(\alpha-1)\pi/2] \tag{3.16}$$

It can be shown that this series is nowhere derivable if $H-\alpha+1 > 1$, i.e. if $H > \alpha$; but is derivable if $H < \alpha$. This is to state that $_{-\infty}\mathscr{D}_t^{\alpha}W(t)$ exists everywhere if $\alpha > H$, and nowhere if $\alpha < H$, which in our notation gives

$$\delta(W) = H \tag{3.17}$$

Put otherwise, this means that there is a relation between the power density spectrum of the graph and the maximum order of (strictly) fractional derivation. It can be shown (cf. Part II) that if Δ is the fractal dimension of the graph, then

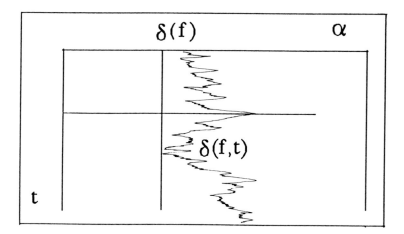

Figure 3.1 Partitioning of the plane, defining the region of derivability.

$H = 2\Delta$, so in fact H is the co-dimension; and we have

$$\delta(W) = 2 - \Delta \qquad (3.18)$$

a remarkable relation which, for a self-affine graph, relates the fractal dimension to the cut in the order of derivation.

COMMENT. If it can be assumed that the spectrum is continuous an integrating factor enters into the calculation of the dimension, changing $1/(i\omega)^{2H}$ to $1/(i\omega)^{2H+1}$ and leading to the result

$$2H + 1 = 5 - 2\Delta \qquad (3.19)$$

Notice that the greater the value of H, the smaller the value of Δ and therefore the smaller the phase shift $(\alpha - 1)\pi/2$ that results from the derivation, and also the spread of the energy density spectrum. It seems to us that these properties must have a deep physical significance. Thus H is the exponent of a geometrical transfer function $C(\omega) \sim 1/i\omega^{2-\Delta}$ imposed by the parametrisation of the curve on the dynamic state function at the fractal boundary (c.f. §2.4). To say that H maximises the order of derivation is equivalent to saying that Δ is minimised so that the real, or "resistive", part of the transfer function $C(\omega)$ is also minimised. This result, valid for a self-affine graph such as that of the Weierstrasse function, can be extended to self-similar distributions, for which it has been shown that $C(\omega) \sim 1/i\omega^{1/\Delta}$. Thus we may conjecture that for such distributions

$$H = 1/\Delta \qquad (3.20)$$

Having the main aim of stimulating interest in this fascinating subject among non-mathematicians we have avoided going into great detail in the discussions; a deeper analysis can be found in the writings of C. Tricot, to which references are given in the Bibliography, and these and other references given provide all the mathematical apparatus needed for the study. It must be emphasised that many problems remain unsolved; whilst these are not necessarily new problems, the links we have attempted to establish between the various concepts of approximation, dimension, transformation, smoothing and non-integral derivation do represent a new approach and may help in taking a new view of the physical problems of heterogeneous media.

3.4 "Fractal" derivatives, Nottale's derivatives, projective derivatives

The concept of fractional derivation arises quite naturally when the parametrisation of the fractal curve is intrinsic: this follows from the work of the present author and of Tricot, but also from that of Nottale who, however, did not use the fractional derivation operator. Nottale's interests concerned relativity and led him to develop a type of derivative that could be called "projective" and which he himself called "fractal".

The new concept of fractal derivation which we now define arises from the relation between fractal and classical co-ordinates. Our view is that classical co-ordinates are obtained when fractal space-time is smoothed by means of balls of radii greater than some critical value. Therefore the classical variables should be identified with (some of the) x_i's along which fractal space-time extends; the others have no direct classical spatio-temporal meaning, along them the fractal space-time extension is of the order of the critical value, which Nottale identifies with de Broglie length and time.

We are now led to investigate the fractal structure of the subset of a fractal included in an infinitesimal subset of the space E in which it is embedded, and in particular in a subset defined by infinitesimal intervals of the variables that are to be identified with the classical ones. Consider the subset included in the domain $(x_i + dx_i)$. The intersection of a "regular" fractal $F(x_i)$ of dimension Δ with a $d\tau$ – manifold is another fractal $F'(x_i)$, the "fractal derivative" of $F(x_i)$, of dimension $\Delta - d\tau$. We can define a "fractal content" $\Lambda_0{}^\Delta(x_i)$ for F' and the infinite hyper-volume is given by

$$v' = \Lambda_0{}^\Delta(x_i)[q^\kappa]^{\Delta - 1 - (d\tau - 1)} \tag{3.21}$$

so that in the Mandelbrot equation the frequency p^k can be written (see Figure)

$$p^\kappa = \Lambda_0{}^\Delta(x_i)[q^\kappa]^\Delta = v'[q^\kappa]^{d\tau} = \Lambda_0{}^\Delta(x_i)[q^\kappa]^{\Delta - 1} = v'[q^\kappa]^{d\tau - 1} \tag{3.22}$$

We now take $dx_i = q^{-k}$ where F is defined as a standard part of a given set F_k, i.e. $F = \text{st}(F_k)$, made up of $\Lambda_0{}^\Delta q^{\kappa\Delta}$ elementary infinitesimal hypervolumes q^{-kd}. *Each of these elementary hypervolumes can be replaced by a single point* giving a set of non-standard points P_k, and the properties of non-standard analysis ensure that the standard fractal F is still given by $F = \text{st}(P_k)$. Equation 3.21 can now be understood as meaning that in the interval $(x_i, x_i + q^{-k})$ there are $\Lambda_0{}^\Delta(x_i)q^{\kappa(\Delta - 1)}$ points of P_k.

The hypervolume of the subset of F included within $(x_i, x_i + dx_i)$ is thus

$v'q^{-k}$, so F can be reconstructed by summing these subsets between the appropriate limits:

$$V = \Sigma v' q^{-\kappa} = \Sigma \Lambda_0^\Delta(x_i)q^{\kappa(\Delta - d\tau)}q^{-\kappa} = q^{\kappa(\Delta - d\tau)}\Sigma \Lambda_0^\Delta(x_i)q^{-\kappa} \tag{3.23}$$

Here $\Sigma \Lambda_0^\Delta(x_i)q^{-\kappa}$ is the non-standard equivalent of the traditional integral $\int \Lambda_0^\Delta(x_i)\,dx$ and if

$$V = \lambda_0^\Delta(x_i)q^{\kappa(\Delta - d\tau)} \tag{3.24}$$

we find

$$\lambda_0^\Delta = \int \Lambda_0^\Delta(x)\,dx \tag{3.25}$$

F is obtained from F' by integration, so inversely we can define F' as the derivative of F with respect to x_i; we call this the *projective derivative*, or *Nottale derivative*, and write it ∂_x

$$v' = \partial_x(V) = \partial_x(\lambda_0^\Delta)q^{\kappa(\Delta - 1) - (d\tau - 1)} = \partial_x(\lambda_0^\Delta)q^{\kappa(\Delta - d\tau)} \tag{3.26}$$

The exponent is again $\Delta - D\tau$ since both the topological and the fractal dimensions have been decreased by 1. In other words, the projective derivative is simply the ratio to dx of the sum of all the hypervolume elements included in $(x, x + dx)$:

$$\partial_x(V) = (\Sigma_i dV_i)/dx \tag{3.27}$$

We now study the relation between the internal variable s, obtained by parametrisation, and the external variable x. As shown in Chapter 2 (see Fig. 2.5)

$$x(s) = \Sigma p_{sk}\cos\varphi_{sk}q - k \tag{3.28}$$

in which form the relation between fractal and classical variable is independent of the embedding space. The fractal derivate can be computed from the equation

$$x \leqslant \Sigma p_{sk}\cos\varphi_{sk}q - k \leqslant x + q - k \tag{3.29}$$

which defines a set $\mathcal{D}$ that contains an infinity of elements, the (infinite) number of points of the non-standard fractal F_k included in the interval $(x, x + q^{-k})$. The non-standard infinite number

$$\partial_x s = [\lambda_0{}^\Delta(x_i)]_\kappa q^{\kappa(\Delta-1)} = \text{Card } \mathscr{D}_\kappa(x) \tag{3.30}$$

is what Nottale calls the fractal derivative, and is also called the projective derivative.

The projective derivative can be defined equivalently as the sum of the infinitesimal sections ds_i of the curvilinear co-ordinate s:

$$\partial_x s = \Sigma ds_i/dx \tag{3.31}$$

For any approximation F_k to the fractal curve corresponding to a resolution $\Delta x = q^{-k}$, the number of solutions becomes finite and proportional to $q^{-k(\Delta-1)}$. The slope of the curve remains undefined, since it is a multi-valued variable; this leads naturally to a probabilistic description.

Chapter 4

Composition of fractal geometries

Physical phenomena can be classified in at least two ways:

(a) Those that are governed by scaling laws related solely to the fractal dimension of a support for a dissipative process (cf. Chapter 2). Every experiment whose objective is a measurement relies on some form of dissipation, and in fact experimentation without dissipation is impossible. In this case the seat of the dissipation is a set whose dimension is either the fractal dimension itself, typically the process of δ-transfer and the TEISI model, or the co-dimension (boundary conditions determined by the fractal limits, as shown in the TEISA model). For this class the measure is said to be homogeneous.

(b) Those whose parameters depend on a combination of several indices, only some of which have the geometrical meaning of a fractal dimension in accordance with the definition given in Chapters 1 and 2. An example is diffusion in a fractal set. The measure here is said to be heterogenous.

In the preceding chapters we have defined fractal geometries and fractal dimensions, shown how these dimensions can be measured and have discussed the time required to perform the measurements; but we have said nothing about the physics of any phenomena that can occur in such geometries. The definition of a parametrisation can be regarded as the implied use of a particular physical process, since it enables us to specify the movement of an arbitrary object on the fractal even though its instantaneous "speed" is formally infinite. We have shown that, given the parametrisation, a variable speed of propagation can be assigned and hence the distribution of time along the curve and all the scaling laws associated with that parametrisation. The variable speed distribution leads to a mixing of information of different types and to some confusing of the transfer function or state function at the interface that has to be taken into account in all linear dissipative processes that involve the geometry. The relations between characteristic lengths and times would clearly be affected and the concept of a unique fractal dimension would have lost its physical significance. It is possible to assign a variable measure to a fractal support: this is the fundamental reason for the need to go beyond the concept of the simple fractality that we have been developing and has led to the extension to *multifractality*.

Mandelbrot used this extended concept even before he invented the term fractal, in order to describe the complex and irregular phenomena of turbulence (1974). It arises quite naturally in the analysis of phenomena such as, for example, the dissipation of energy in random electrical networks (cf. § 8.4); in the context of turbulence, among others, it was rediscovered by Frisch *et al* in 1985.

In this chapter we undertake a very elementary study of multifractality, in the Mandelbrot sense, and then look into the existence of other mixtures of scale laws based on fractional derivation: the aim of this is to account for irreversibility in a fractal medium. In all this we keep to the Mandelbrot meaning of the term multifractality; we need to distinguish it, at least temporarily, from the property of an object whose transfer or state function is characterised by a mixture of derivatives of fractional order, so for this we have used the term *hyperfractality*.

4.1 Statistical aspects: multifractality, the composition of several fractal dimensions

4.1.1 *Distortion of the scaling laws*

Consider again the Cantor set, but now let the original bar have a mass, uniformly distributed over its length, which we can take as unity; and suppose that this mass is conserved in the dissection process (see Fig. 4.1). At the first stage the bar is replaced by two bars of equal lengths $\eta_1 = \eta_2 = 1/3$ and masses, since the total mass is conserved, $\mu_1 = \mu_2 = 1/2$; and therefore of densities $\rho_1 = \rho_2 = 3/2$. At the n'th stage there are $N_n = 2^n$ bars, each of length $\eta_n = 1/3^n$, mass $\mu_n = 1/2^n$ and density $\rho_n = (3/2)^n$. Thus the density increases at each stage, and we have the relation

$$\rho_n = \mu_n/\eta_n = \eta_n^{\alpha-1} \quad \text{where} \quad \alpha = \log 2/\log 3 \tag{4.1}$$

Thus the density tends to infinity according to a power law with exponent $\alpha - 1$, and α is called the *Lipschitz-Hölder exponent*, or the *exponent of the*

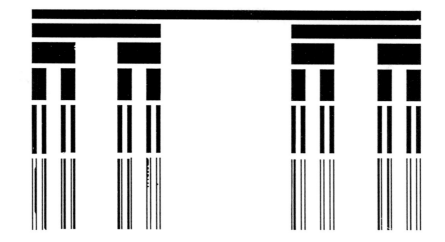

Figure 4.1 Cantor set, supporting a mass distribution that is conserved and remains homogeneous throughout the iteration.

singularity. In this case the exponent α is the fractal dimension Δ of the ultimate Cantor set.

We have described the process in terms of mass, but of course we could have used any other "extensive" property (in the thermodynamic sense of the term) which is conserved. In fact, it is the "μ", however interpreted, that underlines the relevance of a measure for the support.

The mass contained in a set bounded by 0 and x is

$$M(x) = \int_0^x \rho(u)\,du = \int_0^x d\mu \tag{4.2}$$

The graph of $M(x)$ is a "staircase", with steps all of unequal heights – known as the "devil's staircase": see Figure 4.2.

4.1.2 *Multiplicative binomial processes*

Repeated bisection of the interval $[0, 1]$ leads at the n'th stage to the creation of $N_n = 2^n$ segments each of length $\eta_n = 2^{-n}$. Suppose we have a population of N_n individuals distributed over these segments, with segment i, $i = 0, 1, \ldots N_n - 1$,

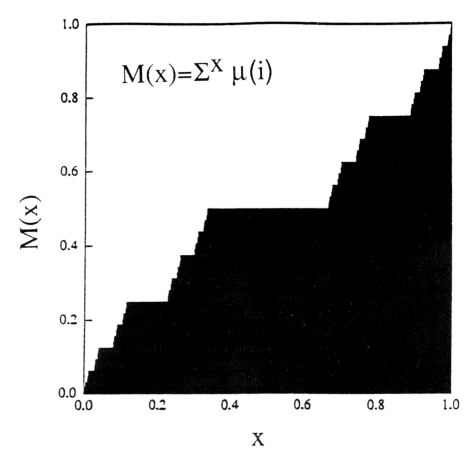

$$M(x)=\Sigma^X \, \mu(i)$$

Figure 4.2 "Devil's staircase": mass distribution for the set of Fig. 4.1.

containing N_i. If we assign a weight $\mu_i = N_i/N_n$ to segment i, this is the probability p_i that a randomly-chosen individual will be in segment i, and the total weight of the distribution is

$$M = \sum_i \mu_i = 1 \tag{4.3}$$

Suppose now that throughout the bisection process the population carried by any element is assigned to one or other of the elements into which it is divided with constant probability, say to the left with probability p and therefore to the right with probability $q = 1 - p$. Thus for the first two stages we have

(1)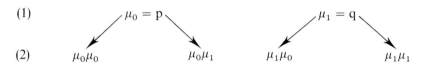

(2) $\quad\quad \mu_0\mu_0 \quad\quad\quad\quad\quad \mu_0\mu_1 \quad\quad\quad \mu_1\mu_0 \quad\quad\quad\quad\quad \mu_1\mu_1$

so the sequences of suffixes are

$$n = 1 \quad\quad\quad\quad\quad 01$$
$$n = 2 \quad\quad\quad\quad\quad 00, 01, 10, 11$$

Continuing, we see that the sequence of 0s and 1s for $n = 3$ is the same as that of the binary fractions $i/2^3$, $i = 0, 1, \ldots 7$:

i	0	1	2	3	4	5	6	7
$i/2^3$	.000	.001	.010	.011	.100	.101	.110	.111

and this obviously generalises to all stages, giving 2^n blocks of n binary digits each at stage n.

At the nth stage the length of the i'th element is $\eta = 2^{-n}$ and the fraction of the population that it carries is given by $\mu_i = \mu_0{}^k \mu_1{}^{n-k}$, where k is the number of zeros in the binary fraction representing the number $i/2^n$.

Let $\theta = k/n$, $k = 0, 1, 2, \ldots n$; then

$$N_n(\theta) = C_n{}^k = n!/k!(n-k)! \tag{4.4}$$

is the number of cells having the measure $\mu_\theta = \Delta^n(\theta)$, where

$$\Delta(\theta) = \mu_0{}^\theta \mu_1{}^{1-\theta} = p^\theta (1-p)^{1-\theta} \tag{4.5}$$

The total weight of the distribution is

$$M(1) = \Sigma\mu_i = \Sigma N_n(\theta)\Delta^n(\theta) = (\mu_0 + \mu_1)^n = 1 \tag{4.6}$$

Figure 4.3(a) shows the probability distribution for the mass after 11 iterations with $p = 0.25$, $1 - p = 0.75$ [Feder 1988].

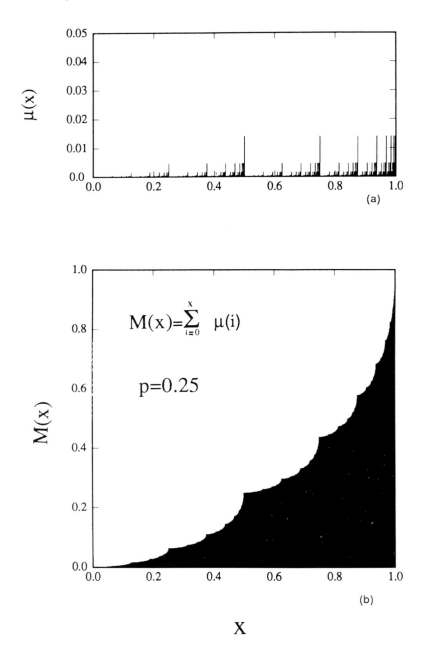

Figure 4.3 (a) Mass distribution after 11 iterations with constant probability p = 0.25 of assignment to the left-hand segment. (b) Cumulative mass distribution for this case.

From this we can calculate the cumulative distribution along the line

$$M(x) = \sum_{0}^{x} \mu_i \qquad (4.7)$$

where $x = i \cdot \eta = i \cdot 2^{-n}$; this is shown in Figure 4.3(b). The calculation shows that

$$\begin{aligned}
M(x) &= p \cdot M(2x) && \text{for} \quad 0 < x \leqslant 1/2 \\
&= p + (1-p) \cdot M(2x-1) && \text{for} \quad 1/2 \leqslant x \leqslant 1
\end{aligned} \qquad (4.8)$$

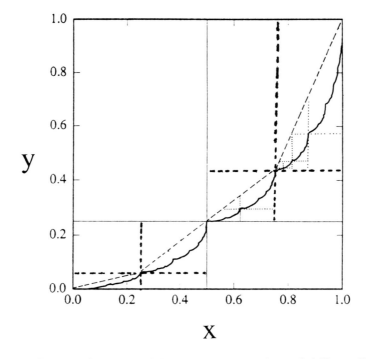

Figure 4.4 Affine transformation of the plane, based on the probability scaling law of Equation 4.8.

Whilst the above transformations mask to some extent the fractal properties of the support for the mass distribution, an affine transformation of the plane into itself can be constructed such that $M(x)$ is obtained after a finite number of transformations (cf. Fig. 4.4). This is:

$$(x, y) \rightarrow (x/2, py) \qquad\qquad [? 0 < x \leqslant 1/2]$$
$$(x, y) \rightarrow (1/2, p) + [1/2, (1-p) \cdot y] \, [?1/2 \leqslant x \leqslant 1] \qquad (4.9)$$

which, starting with the line $y = x$, reproduces exactly the curve $y = M(x)$ of Fig. 4.3(b). A generalisation of this result is given in § 8.5.

4.1.3 *Fractal subset*

At the n'th iteration there are $N_n(\theta)$ elements, each of length η_n, with the same measure (i.e. supporting the same population) μ_θ, where $\theta = k/n$, $k = 0, 1, \dots n$; these define a sub-set, $S_n(\theta)$, say, characterisd by all the members having the same number k of zeros in their representative binary fraction. As $n \rightarrow \infty$, S_n tends to a limiting set, S_θ, say. It is interesting to investigate the possible fractal dimensions of the S_θ, which we should expect to depend on θ and therefore on k. For this, we need to find the Δ-measure of S_θ, defined in Chapter 2 by the cut

$$M_\Delta(S_\theta) = \Sigma \eta^\Delta \quad \text{summed over } S_\theta$$
$$= \lim N_n(\theta) \eta^\Delta \quad \text{as} \quad n \rightarrow \infty, \quad \eta \rightarrow 0$$
$$= 0, \quad \Delta > f(\theta)$$
$$= \infty, \quad \Delta < f(\theta) \qquad (4.10)$$

Using Stirling's formula $n! \sim (2\pi n)^{1/2} n^n e^{-n}$ for large n we find

$$N_n(\theta) \sim [2\pi n\theta(1-\theta)]^{-1/2} \cdot \exp\left(-n[\theta \ln(\theta) + (1-\theta)\ln(1-\theta)]\right) \qquad (4.11)$$

Since $\eta = 2^{-n}$, $n = -\log \eta / \log 2$, and hence

$$M_\Delta(S_\theta) \sim \eta^{D-f(\theta)} \qquad (4.12)$$

where

$$f(\theta) = -[\theta \log(\theta) + (1-\theta)\log(1-\theta)]/\log 2 \qquad (4.13)$$

Thus if $\Delta = f(\theta)$ the measure remains finite and non-zero as $n \rightarrow \infty$, and therefore this is the fractal dimension of the sub-set. The result provides a striking illustration of the concept of dimension as defined by a cut in the real numbers.

The limiting population generated by the process is distributed over a set S which is the union of all the sub-sets S_θ. Since the fractal dimension of each sub-set depends on the relevant value of θ, we have in S an example of *multifractality*, the union of a number of sets each having its own fractal dimension.

4.1.4 *The Lipschitz-Hölder exponent*

It appears in practice that the variable θ is not very convenient to use, and for this reason Mandelbrot at an early stage recommended the use of another, the Lipschitz-Hölder exponent α (see Mandelbrot 1982, p. 373). This is defined as follows.

Let $x(\theta) = i/2^n = \Sigma \varepsilon_v \cdot 2^{-v}$, with all $\varepsilon_v = 0$ or 1, be a point of the set S_θ. We can find the value of the measure $M[x(\theta)]$ and also that at the point $x(\theta) + \eta$, $M[x(\theta) + \eta]$; the increment in going from $x(\theta)$ to $x(\theta) + \eta$ is μ_θ, by definition, which we can represent as a power of η:

$$M[x(\theta) + \eta] - M[x(\theta)] = \mu_\theta \sim \eta^\alpha \tag{4.14}$$

The function $M(x)$ that satisfies this relation for all x has a derivative if $\alpha = 1$, is constant of $\alpha > 1$ and is singular if $0 \leqslant \alpha < 1$. From the expression for μ_θ (equation 4.5) we find

$$\alpha(\theta) = \log \mu_\theta / \log \eta = -[\theta \log(p) + (1 - \theta) \log(1 - p)]/\log 2 \tag{4.15}$$

$\alpha(\theta)$ has the form of a dimension; its values lie in the interval bounded by

$$\alpha(\theta = 0) = -[\log(1 - p)]/\log 2$$
$$\alpha(\theta = 1) = -[\log(p)]/\log 2 \tag{4.16}$$

There is a bijective relation between α and θ, so the function $f(\theta)$ defined by Equation 4.13 can be transformed into a function $f(\alpha)$ of α; this is shown graphically in Figure 4.5. The curve of $f(\alpha)$ has a number of characteristic features, one of which is that it has a maximum at the value of α corresponding to $\theta = 1/2$.

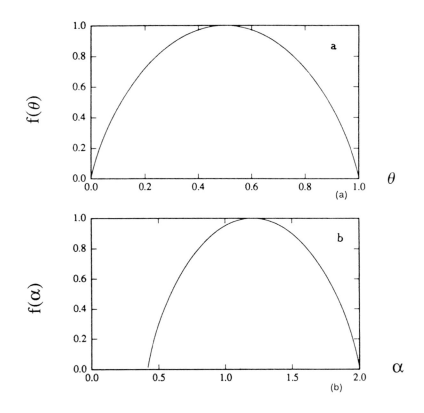

Figure 4.5 Results of a binomial decomposition with probability $p = 0.25$. (a) Fractal dimension of the sub-set S_θ, where θ is determined by the number of zeros in the binary-fraction representation of the x-co-ordinate. (b) Fractal dimension of the sub-set S_α.

4.1.5 *Dimension and entropy*

As we have just shown, the distribution procedure in the limit disperses the original population among sub-sets of different dimensions. At the n'th stage the measure of the sub-set $S_n(\theta)$ is

$$M_n[S_n(\theta)] = N_n(\theta)\mu_\theta$$

$$= [2\pi p(1-p)]^{-1/2} \exp[-n(\theta-p)^2/2p(1-p)] \qquad (4.17)$$

The population is centred on the value $\theta = p$ and decreases as $|\theta - p|$

increases like $n^{-1/2}$, creating a sub-set having a probability distribution about a mean:

$$S_\sigma = \cup \, S_n(\theta)$$

$$p - \sigma \leqslant \theta \leqslant p + \theta$$

$$M(S_\sigma) = [2\pi np(1-p)]^{-1/2} \int_p^{p+\sigma} \exp[-n(\theta-p)^2/2p(1-p)]\,d\theta$$

$$= (2/\sqrt{\pi}) \int_0^\tau \exp(-t^{-2})\,dt \quad \text{where} \quad \tau = \sigma\sqrt{[n/2p(1-p)]}$$

in the limit $n \to \infty$, $\sigma \to 0$. Thus we find that a finite fraction of the mass, arbitrarily close to 100%, is contained in a set S_σ. The dimension of this set is

$$f(\theta) = -[p\log(p) + (1-p)\log(1-p)]/\log 2] \tag{4.18}$$

in which we can recognise the entropy in Shannon's sense of the term. Thus for this process there is a relation between the entropy and the dimension of the most probable set in the multifractal set.

4.1.6 Distribution of the mass exponents $\tau(q)$

We now add a parameter q and complicate the problem by distributing the mass heterogeneously among the discs used in defining the measure:

$$M_\Delta(q,\eta) = \sum_i \mu_i^q \eta^\Delta = N(q,\eta)\eta^\Delta \to 0 \quad \text{if} \quad \Delta > \tau(q)$$

$$\to \infty \quad \text{if} \quad \Delta < \tau(q)$$

Every finite measure requires that

$$N(q,\eta) = \sum_i \mu_i^q \sim \eta^{-\tau(q)} \tag{4.19}$$

so the mass exponent is given by

$$\tau(q) = -\lim_{\eta \to 0} [\log N(q,\eta)/\log \eta] \tag{4.20}$$

For q = 0, $\mu_i^q = 1$ and $N(\eta)$ is the number of discs needed to cover the set; thus dim $S = \tau(0)$. If q = 1, $N(q, \eta) = \Sigma\mu_i = 1$, since the total mass is conserved, so $\tau(1) = 0$.

Consider now the behaviour of the derivative $d\tau/dq$. A short calculation (Feder 1988) shows that as $\eta \to 0$ the values of this as $q \to \pm\infty$ give the bounds for the Lipschitz-Hölder exponent:

$$\alpha_{max} = -\lim(d\tau/dq), \quad q \to -\infty$$

$$\alpha_{min} = -\lim(d\tau/dq), \quad q \to +\infty$$

The interesting case is q = 1: for this

$$(d\tau/dq)_{q=1} = -\lim_{\eta \to 0}[\Sigma(\mu_i \log \mu_i)/\log \eta] = \lim[S(\eta)/\log \eta] \tag{4.21}$$

where $S(\eta)$ is the Shannon entropy for the discs of diameter η. Thus

$$S(\eta) \sim -\alpha_1 \log \eta \tag{4.22}$$

where $\alpha_1 = (d\tau/dq)_{q=1}$ is the fractal dimension of the set in which the measure is concentrated.

We next look for a relation between $\tau(q)$ and the function $f(\alpha)$ defined in §4.1.4 and shown in Figure 4.5. We can write

$$S = \cup_\theta S_\theta = \cup_\alpha S_\alpha \tag{4.23}$$

$$N(\alpha, \eta)\eta^{f(\alpha)} = \rho(\alpha) d\alpha \tag{4.24}$$

where $N(\alpha, \eta)$ is the number of sets between S_α and $S_{\alpha+d\alpha}$.

The measure μ_α in the disc of diameter η depends on the particular law of scale, $\mu_\alpha = \eta^\alpha$ (cf. 4.1), from which we get $\mu_\alpha^q = \eta^{\alpha q}$, and therefore

$$M_\Delta(q, \eta) = \Sigma\mu_i^q\eta^\Delta = \int N(\alpha, \eta)\eta^{\alpha q}\eta^\Delta d\alpha \tag{4.25}$$

$$= \int\rho(\alpha)\eta^{\alpha q - f(\alpha)}\eta^\Delta d\alpha \tag{4.26}$$

The integral is dominated by the maximum of the integrating factor, given by

$$(d/d\alpha)[\alpha q - f(\alpha)] = 0 \tag{4.27}$$

If the solution is $\alpha = \alpha(q)$ we then have

$$M_d(q, \eta) \sim \eta^\varphi, \quad \text{where} \quad \varphi = q\alpha(q) - f\{\alpha(q)\} + \Delta \tag{4.28}$$

For the measure to remain finite as $\eta \to 0$ we must have $\varphi = 0$, i.e. $\Delta = \tau(q) = f\{\alpha(q)\} - q\alpha(q)$. Thus the mass exponent is given in terms of the Lipschitz-Hölder exponent $\alpha(q)$ and the fractal dimension $f\{\alpha(q)\}$ of the supporting set, and these can be found from the equations

$$\alpha(q) = (d/dq)\tau(q) \tag{4.29}$$

$$f\{\alpha(q)\} = q\alpha(q) + \tau(q) \tag{4.30}$$

These equations are the Legendre transforms into f and α of the independent variables τ and q. The transforms are not neutral with respect to phenomenological thermodynamics in particular, and Mandelbrot (1989) has shown that a correspondence can be established between q (as the inverse of a temperature T) and $\tau(q)$ (the negative of a free energy), and between $f(\alpha)$ and the Gibbs free energy. A study of Mandelbrot's publications will reveal the rich thermodynamic content hidden in the concept of multifractality.

The importance of the analysis we have just given is that it shows that there is value (cf. Epilogue) in complicating the concept of fractality in this way. However, this makes it more and more difficult to derive the simple concepts which the engineer needs for his daily tasks; and it is with the aim of relieving this difficulty, which arises especially in the study of dissipation in a multifractal medium, that we have introduced the further concept of *hyperfractality*. We now explain this, but first we consider fractional Brownian motion, treated by means of fractional derivation.

4.2 Hyperfractality: composition of derivations of different orders

4.2.1 *From Brownian motion to fractional Brownian motion*

Brownian motion is important from two points of view: historical – it was the first example of a physical process characterised by non-differentiability; and experimental – it provided the basis for the diffusive process, considered as the archetype of dissipative limitation in physics and chemistry.

To establish the background, consider a particle moving at random along the x-axis, making a jump of length η to the right or the left, with probability

$p(\eta, \tau)$, during each unit of time τ. The variance of this process is $\langle n^2 \rangle = \int \eta^2 p(\eta, \tau)\, d\eta$, and if the probability distribution is Gaussian, i.e.

$$p(\eta, \tau) = (4\pi D\tau)^{-1/2} \exp(-\eta^2/4D\tau) \tag{4.31}$$

this is constant: $\langle \eta^2 \rangle = 2D\tau$. The parameter D is called the *diffusion constant*, and has the dimensions $[L^2 T^{-1}]$, "area per unit time" – see Figure 4.6. The process is normalised by changing the space variable to $\eta/\sqrt{2D\tau}$, giving $\langle \eta^2 \rangle = 1$.

Suppose that at time t the particle is at x(t), so the distance from its (arbitrary) origin is $B(t) = x(t) - x(0)$. The graph of B(t) as a function of t oscillates with extreme irregularity and has properties very similar to those of the Weierstrasse non-differentiable function, in particular those of scale that are associated with self-affineness. To prove this we use the convolution operator to establish the relation btween $B(\tau)$ and $B(b\tau)$.

The interval ξ can be divided into two independent increments η_1, η_2; the probability of finding the particle in ξ at time τ is given by the convolution

$$p(\xi, b\tau) = p(\xi - \eta_2, t - b\tau - t) * p(\eta_2, t) \tag{4.32}$$

Further, assuming the two intermediate processes are Gaussian,

$$p(\xi, b\tau) = (4\pi D_{iff} b\tau)^{-1/2} \exp(-\xi^{-2}/4 D_{iff} b\tau) \tag{4.33}$$

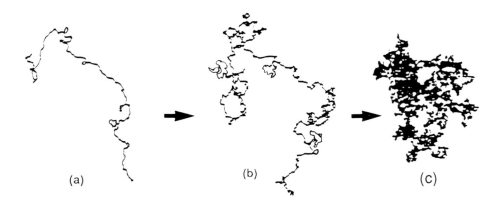

(a)   (b)   (c)

Figure 4.6 Examples of Brownian motion trajectories. (a) $\Delta = 1.1$ (b) $\Delta = 1.45$ (c) $\Delta = 2$. ($\Delta = 1/H$).

(see §5.1 for the physical meaning of the convolution; here D_{iff} is the diffusion coefficient, the more usual notation).

The variance of this process is $\langle \xi^2 \rangle = 2 D_{iff} b\tau$; so the Brownian process is conservative in the sense that the original variance $\langle \eta^2 \rangle$ can be recovered by making an affine transformation, as follows:

$$b\tau \to \tau, \quad \xi \to b^{1/2}\eta \quad D_{iff} \to D_{iff}$$

giving

$$p(b^{1/2}\eta, b\tau) = b^{-1/2} p(\eta, \tau) \tag{4.34}$$

where the factor $b^{-1/2}$ normalises the probability.

Consider next the function $B(t)$, for some given Brownian process, as a random function of time t; for all $t > t_0$

$$\langle [B(t) - B(t_0)]^2 \rangle = (2 D_{iff} \tau) [(t - t_0)/\tau]^{2H}$$
$$\sim [(t - t_0)/\tau]^{2H} \tag{4.35}$$

provided $H = 1/2$. After Mandelbrot, we call any process for which $H \neq 1/2$ a *factional Brownian process*. A point of great importance here, strongly emphasised by Feder (1988), is the strongly auto-correlated nature of such a process. To show this, consider the increment $B_H(0) - B_H(-t)$; the conditional probability of the increment $B_H(t) - B_H(0)$, given the previous increment $B_H(0) - B_H(-t)$, is

$$Pr = \langle [B_H(0) - B_H(-t)] \cdot [B_H(t) - B_H(0)] \rangle \tag{4.36}$$

To simplify the calculation, let $B_H(0) = 0$, $\tau = 1$, $2D\tau = 1$; we then find for the correlation function $C(t)$ between past and future, normalised by $\langle B_H(t)^2 \rangle$,

$$C(t) = \langle [B_H(-t)] \cdot [B_H(t)] \rangle / \langle B_H(t)^2 \rangle = 2^{2H-1} - 1 \tag{4.37}$$

Thus if $H = 1/2$, $C(t) = 0$ and there is no correlation, which is the case for Brownian motion; but if $H \neq 1/2$, $C(t) \neq 0$ and there is a correlation between past and future, and this is independent of the time t. More precisely, since for $H > 1/2$, $C(t)$ is positive, any tendency at a time t_0 for the increment to increase (or decrease) will lead on average to an increase (or decrease) at a later time t; and for $H < 1/2$ the correlation is negative and the effect is inverted – an initial increase results in a later average decrease, and vice versa.

This is a striking result, for it runs counter to the classical state of affairs in which any correlation effects tend on average to disappear – this is the equilibrium hypothesis, or in physics the increase of entropy. What we have shown is that in the appropriate geometry correlations can be conserved.

The fractional Brownian motion can be expressed as a convolution of a standard motion ($H = 1/2$) with a test function $K(t)$ which expresses the scaling law: for given $B_H(0)$

$$B_H(t) - B_H(0) \sim \int_{-\infty}^{t} K(t-t')\,dB(t') = K(t) * dB(t) \tag{4.38}$$

where

$$K(t-t') = (t-t')^{H-1/2}, \qquad\qquad 0 \leqslant t' \leqslant t$$
$$= (t-t')^{H-1/2} - (-t')^{H-1/2}, \quad t' \leqslant 0$$

This representation was proposed by Mandelbrot and van Ness in 1968; it can be expressed also in terms of fractional derivation, as follows.

For $0 \leqslant t' \leqslant t$

$$B_H(t) - B_H(0) = {}_0\mathscr{D}^{-H+1/2} B(t) \tag{4.39}$$

The scaling laws can be investigated by putting

$$B_H(bt) - B_H(0) \sim \int_0^{bt} (bt-t')\,dBt' \tag{4.40}$$

Changing the variable of integration to t'', with $t' = bt''$,

$$dB(bt'') = b^{1/2}\,dB(t'') \tag{4.41}$$

$$K(bt-bt'') = b^{H-1/2} K(t-t'') \tag{4.42}$$

hence

$$B_H(bt) - B_H(0) = b^H [B_H(t) - B_H(0)] \tag{4.43}$$

If now we put $t = 1$, $\Delta t = bt$, $B_H(1) - B_H(0) = \delta_t$,

$$B_H(\Delta t) - B_H(0) = |\Delta t|^H \delta_t \tag{4.44}$$

What is of greatest interest to us here is the result that, compared to the classical uncorrelated Brownian motion, fractional Brownian motion with $H > 1/2$ is persistent and tends to propagate itself further, and therefore faster, than the classical diffusive motion. The variance is not now constant but is given by

$$\langle [B(\Delta t) - B(0)]^2 \rangle = (2D_{iff}\tau)[\Delta t/\tau]^{2H} \tag{4.45}$$

Unfamiliarity with motion of this type may lead us to try to represent it as a form of the classical motion; if the above variance (equn. 4.45) is $\xi^2(\Delta t)$ then

$$\xi(\Delta t) = (2D_{iff}\tau)^{1/2}[\Delta t/\tau]^H = 2D_H\sqrt{(\Delta t)} \tag{4.46}$$

where D_H is the "apparent" diffusion coefficient, which now varies with time: $D_H = (2D_{iff}\tau)^{1/2}[1/\tau]^{2H}(\Delta t)^{H-1/2}$.

This diffusion can be categorised as abnormal even though we have, in the Laplace space and with the notation of Table 2.1

$$\xi(\Delta t) \sim (\Delta t)^H \rightarrow \eta(p) \sim p^{-H} \tag{4.47}$$

or in terms of fractional derivatives

$$\mathscr{D}_t^{H-1}[\xi(t)] \sim \delta_t, \quad \mathscr{D}_t^H[l(t)] \sim \delta_t \tag{4.48}$$

On other words, fractional Brownian motion can be defined by means of a measure, one possibility being the time derivative of non-integral order H of the distance of the distance $l(t)$ travelled by the particle executing the motion. With this approach the appearance of a diffusion coefficient that varies with time is simply an artefact of the calculation, with no deep physical significance.

This result is one of the most useful products of the analysis because, as the TEISI model shows, it gives the engineer – who is in need of invariants – the possibility of using characterisation parameters which are also operational variables.

4.2.2 *Composition of fractional derivation indices: hyperfractality*

In the method just given we have introduced a further parameter in order to account for fractional motion; this is typical of many methods of attack on scientific problems, but it is well known that great care must be taken in adding parameters, for otherwise the whole interest of the problem can be destroyed. To complete our analysis with this in mind we shall now study three physical situations characterised by the composition in time of different fractal dimensions.

Returning to the natural parametrisation of a self-similar curve (Table 2.1), we have

$$l(t) \sim t^{(1/\Delta)-1} \tag{4.49}$$

We have stated that this relation is simply the combination of two others which have different physical bases:

$$l(t) \sim \xi^{1-\Delta} \tag{4.50}$$

which expresses the length of a fractal curve, and

$$t \sim \xi^{\Delta} \tag{4.51}$$

which expresses the parametrisation of the curve if it is self-similar: for a self-affine curve we should have

$$t \sim \xi \tag{4.52}$$

which would lead to

$$l(t) \sim t^{1-\Delta} \tag{4.53}$$

The analysis of the preceding chapters has led to relations between the natural parametrisations, the fractional order of derivation that gives the measure of the geometry (the fluxes), the test function (the flux densities) and the fractal dimension of the boundary curve. We shall now show how the degree of freedom provided by the nature of the mapping γ can lead to a compounding of the fractal dimensions that are brought into play in the course of complex physical processes that involve more than one geometry.

The spectral dimension

This concept was introduced by Alexander and Orbach (1982) and developed by Rammal and his collaborators (1983), to account for the propagation of acoustic modes in fractal structures. It can be understood very easily as a departure from ordinary diffusion that makes no specific appeal to fractional Brownian motion, as follows.

A particle in random motion on a plane network has a probability p(t) of returning to the origin at time t; in Brownian motion $p(t) \sim t^{-1/2}$. Since the value 2 in 1/2 is the dimension of the Brownian trajectory, this suggests a generalisation to

$$p(t) \sim t^{-ds/d} \tag{4.54}$$

where d is the dimension of the space and ds is what we call the spectral dimension.

This probability law can be expressed in another way, by stating that ξ^2, the square of the mean element traversed by the particle – which measures the area of a disc radius ξ – is 1/p: i.e.

$$[\xi(t)]^2 \sim t^{ds/d} \tag{4.55}$$

from which it follows that, unlike the result given by equations 4.51 and 4.52, the mean element traversed is

$$\xi(t) \sim t^{ds/2d} \tag{4.56}$$

What has changed is the mapping.

By analogy with classical diffusion for which $[\xi(t)]^2 \sim t$ we can write

$$[\xi(t)]^{dw} \sim t, \quad \text{where} \quad dw = 2d/ds \tag{4.57}$$

where dw is called the *apparent dimension* of the motion of the particle on the fractal. This is an index associated with a dynamic behaviour which can be expressed as the *time dependence of the diffusion coefficient*. As we saw in §4.2.1, studies of *clusters* in percolation lead very generally to a value in the region of 4/3 for ds, whatever the dimension of the space. This invariance with respect to dimension derives from quantum mechanics, as does the physical significance of ds; which leads us to suspect that the co-dimension (§1.5) may be a factor here.

Following the argument that dissipative phenomena in physics are most often accounted for by random motions, the spectral dimension is being proposed as a way of taking into account the limitations of generalised diffusion theory applied to chemical reactions in complex media. The present author prefers an approach that is both less arbitrary and more predictive, at least for the fields in which his own group has been able to experiment.

Hyper-scale indices

We saw in Chapter 3 that dynamic exchange across a self-similar fractal surface can be expressed by means of a fractional derivative which minorises the spectral density of the power dissipated:

$$\mathscr{D}_t^{(1/\Delta)-1} J(t) \sim \varphi(t) \tag{4.58}$$

where the test function $\varphi(t)$ is the flux-exchange density.

Assuming that the flux density if proportional to the driving force we have $\Delta X(t) \sim \varphi(t)$ and

$$\mathscr{D}_t^{(1/\Delta)-1} J(t) \sim \Delta X(t) \tag{4.59}$$

which is a generalised flux-force relation.

If $\Delta = 2$ this gives the equation for diffusion in a semi-infinite Euclidean medium; what we want now to do is to translate this type of limitation to a fractal medium, *while conserving the natural parametrisation*. We can solve this problem by applying the fractional derivation operator not only to the total flux $J(t)$ but also to the flux density $\varphi(t)$; from equation 4.58 we have

$$\mathscr{D}_t^{(1/\Delta)-1} J(t) \sim \varphi(t) \quad \text{with, for example,} \quad \mathscr{D}_t^{-1/2} \varphi(t) \sim \Delta X(t)$$

The expression for the flux density is given by equation 4.59, from which we get the dynamics of diffusion across a *fractal interface*:

$$\mathscr{D}_t^{(1/dw)-1} J(t) \sim \Delta X(t) \quad \text{with} \quad dw = 2\Delta/(2-\Delta) \tag{4.60}$$

i.e. $1/dw = 1/\Delta - 1/2 = 1/2 + 1/\Delta - 1$

This relation has been broadly verified, in particular for electrochemical dynamics in a porous medium; it can be extended to the case of Brownian motion of order β, characterised by

$$\mathscr{D}_t^{1-1/\beta} \Delta X(t) \sim \varphi(t) \tag{4.61}$$

when equation 4.60 applies with

$$1/dw = 1/\beta + 1/\Delta - 1 \tag{4.62}$$

The importance of this result is that it gives a means for accessing the factors Δ and β independently, and thus for checking experimentally the validity of relations of this type (cf. §7.9–10).

Returning to the case of diffusion in a fractal medium, we have shown that the dynamics is that of δ-transfer with $dw = 2\Delta/(2-\Delta)$ (equation 4.60). The relevant fractal co-dimension is $\zeta = 2-\Delta$, in terms of which $dw = 2\Delta/\zeta = 2(2-\zeta)/\zeta$, identical in form to the relation of equation 4.57 giving dw in terms of the spectral dimension, $dw = 2d/ds$. For the dynamics to correspond to a linear relation between flux and force the value of dw must be 1, and therefore ζ must be 4/3. Thus the necessary and sufficient for this to be the case is simply $\zeta = 4/3$, independent of the dimension of the space provided only that the medium is regular.

"Janals"

We now make full use of the arbitrariness of the mapping γ for parametrising the self-similar curve of fractal dimension Δ by replacing the relation $l(t) \sim N \cdot \xi(t)$ by $l(t) \sim n \cdot [\xi(t)]^D$, while retaining the relation $N \sim t^{-1}$. Here D is an index that measures the fractal content of the curve, differing from the dimension Δ which parametrises the curve with $\xi(t) \sim t^{1/\Delta}$.

We now have

$$l(t) \sim t^{(d/\Delta)-1} \tag{4.63}$$

Put $d/\Delta = da$, the *apparent dimension* of the dynamics. The arbitrariness of the choice of D would be of no interest were it not for the fact that this has remarkable theoretical and practical consequences, which are illustrated by the following example.

Suppose $da = 1/2$, and that this results from taking $D = 2$ and $\Delta = 1$. Then $D/\Delta = 2$, $l(t) \sim t$ and

$$\mathscr{D}_t^{1} J(t) \sim \varphi(t) \tag{4.64}$$

If it can be assumed that the linear character of the local transfer can be preserved we can write $\Delta X(t) \sim \varphi(t)$, and so

$$\mathscr{D}_t^{1} J(t) \sim \Delta X(t) \tag{4.65}$$

The flux and the force which express the dynamics are in quadrature; thus *the system having this double dimensionality is inertial.* The geometry and the parametrisation have led to a system whose dynamics is no longer dissipative: time and space have been linked in such a way that, by combining the dimensions, the dissipative term has vanished. Objects with such a property have been conceived and called *Janals* – from the combination of Janus and Fractal. They are being studied in research on superconductivity (cf. Part II).

This example illustrates the potentialities of fractals: the engineer who gains a mastery of parametrisation in fractal geometry will find that this opens up valuable possibilities for him.

Bibliography

Chapter 1

CEA, EDF, INRIA, *Les Fractales*, Ecole d'Hiver, 1987.

DUBUC S., "Atelier de Géométrie Fractale", *Annales des Sciences Mathématiques du Québec*, 1987.

FEDER J., *Fractals*, Plenum, New York, 1988.

HATA M., "Fractal Mathematics" in *Patterns and Waves: Qualitative Analysis of Non Linear Differential Equation*, 1986, 259.

LAUWERIER H., *Fractals*, Aramith, Amsterdam, 1987.

MANDELBROT B., *Les Fractales*, Flammarion, Paris, 1989.

MANDELBROT B., *The Fractal Geometry of Nature*, Freeman, San Francisco, 1982.

MANDELBROT B., *Fractals and Multifractals: Noises, Turbulence and Galaxies*, Selecta, vol. 1, Springer-Verlag, Berlin, 1990.

PEITGEN H.O. and RICHTER P.H., *The Beauty of Fractal*, Springer-Verlag, Berlin, 1986.

PEITGEN H.O. and SAUPE D. (Eds.), *The Science of Fractal Image*, Springer-Verlag, Berlin, 1988.

SAPOVAL B., *Les Fractales*, Aditech, Paris, 1989.

STEWART I., *Les Fractales* (*Les chroniques de Rose Polymath*), French comic strip, Eds. Belin, Paris, 1983.

VOSS R.F., "Random Fractal Forgeries" in *Fundamental Algorithms for Computer Graphics*, NATO ASI, vol F 17, edited by R.A. Earnshow, Springer-Verlag, New York, 1985.

VOSS R.F., "Random Fractals: Characterisation and Measurement" in *Scaling Phenomena in Disordered Systems*, edited by R. Lynn and A. Skjeltorp, Plenum, New York, 1985.

Programmes and video

"Nothing but Zooms", Art Matrix, P.O. Box 880, Ithaca, NY 14851 – 0880; tel: (607) 277-0959. 3 minutes.

"Fractal Fantasy", C. Fitch (Media Magic, 1987). Media Magic, P.O. Box 2069, Mill Valley, CA 94942; tel: (415) 381-4224.

"Aggregation", M. Kolb (ZEAM, Frei Universitat Berlin, 1985). 25 minutes. M. Kolb, Laboratoire de Physique des Solides, Bat. 510, Université de Paris sud, 91405 Orsay, France.

"Introduction to Fractals", L. Fogg, Micro Cornucopia (USA), n° 33, 36-40 (Dec. 1986/Jan. 1987).

"Fractals", P.R. Sorensen, BYTE (USA) 9, 157-172 (1984). Programme BASIC.

"Gold DLAs", A.J. Hurd and D.G. Madeleine (Sandia National Laboratories, Albuquerque, 1987). 3 minutes.

"Exploring Mandelbrot Fractals with MMSFORTH", J.A. Miller and J.J. Gerow, J. Forth Appl. Res. 4, (1986), 339-42.

"Generation and Display of Geometric Fractals in 3D", V.A. Norton, Comput. Graph. 16, (1982) 61-5.

"From Here to Reality (Computer Graphics Animation Systems)", S. Beesley, Your Comput. (GB) 4, 1984) 82-83.

"Random Fractal Forgeries", R.F. Voss, in *Physics-Like Models of Computation*, Proc. NATO, Adv. Study Inst. on Fundamental Algorithms in Computer Graphics (Ilkley, Yorkshire, UK, 1985); *Physica* 10D (1984) 81.

Chapter 2

Dubuc B., Quiniou J.F., Roques-Carmes C., Tricot C. and Zucker S.W., McGill Research Center for Intelligent Machines T.R. CIM 87.15 1987; *Visual Communication and Image Processing*, 11, 1987, 241.

Falconer K.J., *The Geometry of Fractal Set*, Cambridge Tracts in Mathematics, Cambridge University Press, 1986.

Kaye B., *A Random Walk Through Fractal Dimension*, VCH, New York, 1989.

Koto N., "On Self Affine Functions", *Japan J. Appl. Math. 3*, 1986, 259.

Le Méhauté A., "Introduction to Transfer and Motion in Fractal Media: the Geometry of the Kinetics", *Sol. Sta. Ionics 9&10*, 1983, 17.

Tricot C., "Rarefaction Indices", *Mathematika 27*, 1980, 46.

Tricot C., *Mesures et Dimensions*, doctoral thesis, Université de Paris Sud, Orsay, 1983.

TRICOT C., "Les Avatars Successifs de la Dimension Fractale", *Gazettes des Sc. Math. du Québec* 10(2), 1986.

TRICOT C., "Dimensions des Graphes", *C.R. Acad. Sc.*, Paris, 303, 1986, 609.

TRICOT C., QUINIOU J.F., WHEBI D., ROQUES-CARMES C. and DUBUC B., "Evaluation de la Dimension Fractale d'un Graphe", *Rev. Phys. Appl.* 23, 1988, 111.

TRICOT C., "Dimension Fractale et Spectre", *J. Chim. Phys. 85*, 1988, 379.

TRICOT C., "Dérivation Fractionnaire et Dimension Fractale de son Graphe", *Rapport 1532 du C.R. Mathématique*, Institut Polytechnique de Montréal, 1988.

TRICOT C. and LE MÉHAUTÉ A., *Mesure de frontières et flux*, in press.

Chapter 3

HATA M., "Fractal Mathematics" in *Patterns and Waves: Qualitative Analysis of Non Linear Differential Equation*, 1986, 259.

OLDHAM K.B. and SPANIER J., *The Fractional Calculus*, Academic Press, New York, 1974.

OUSTALOUP A., *Systèmes Asservis Linéaires d'Ordre Fractionnaire: Théorie et Pratique*, Masson, Paris, 1983.

ROSS B., *Fractional calculus and its Applications*, Springer-Verlag, Berlin, 1974.

TRICOT C., "Dérivation Fractionnaire et Dimension Fractale de son Graphe", *Rapport 1532 du C.R. Mathématique*, Institut Polytechnique de Montréal, 1988.

WEHBI D., *Approche Fractale de la Rugosité de Surface et Applications Analytiques*, doctoral thesis, Besançon, 1986.

Chapter 4

Multifractality

FRISCH U. and PARISI G., "A Multifractal Model of Intermittency" in *Turbulence and Predictability in Geophysical Fluid Dynamics and Climate Dynamics*, edited by A. Ghil, P. Benzi and G. Parisi, North Holland, Amsterdam, 1985.

GABRIEL P.S., LOVEJOY D. and SCHERTZER D., *J. Geophys. Res. Lett.*, 1988, 1373.

KAHANE J.P., *Ann. Sci. Math. Que. 9*, 1988, 435.

LAVALÉE D., SCHERTZER D. and LOVEJOY S., "On the determination of the codimension function" in *Scaling Fractal and Non Linear Variability in Geophysics*, edited by D. Schertzer and S. Lovejoy, in press.

LOVEJOY S. and SCHERTZER D., *Wat Resour. Res. 21*, 1985, 1233.

LOVEJOY S. and SCHERTZER D., *Wat. Resour. Res. 25*, 1985, 577.

LOVEJOY S. and SCHERTZER D., *Bull AMS 67*, 1986, 21.

LOVEJOY S. and SCHERTZER D., TSONIS A.A., *Science 235*, 1987, 1036.

MANDELBROT B., "Limit in Normal Multifractal Measures" in *Frontiers of Physics: Landau Memorial Conference*, edited by E. Gotsman, Pergamon, New York, 1989.

MANDELBROT B., "Negative Fractal Dimension and Multifractals" in *Statphys. 17*, edited by C. Tsallis, North Holland, Amsterdam, 1989.

MANDELBROT B., "A Class of Multinomimal Multifractal Measures with Negative (Latent) Value of the 'Dimension' $f(\alpha)$" in *Fractal*, edited by L. Pietronero, Plenum, New York, 1989.

MANDELBROT B., *Multifractal Measures, Especially for Geophysicics*, Pageoph. 131 (1/2), 1989, 133.

MENEVEAU C. and SREENIVASAR K., *Phys. Rev. Lett. 59 (13)*, 1987, 1424.

SCHERTZER D. and LOVEJOY S., *Ann. Math. Que. 11*, 1987, 139.

STANDLEY E. and MEAKING P., *Nature 335, 1988*, 405.

The Spectral dimension

ALEXANDER S. and ORBACH R., "Density of states on fractals: 'fractons'", *J. de Phys. (Paris) lett 43*, 1982, L625.

ALEXANDER S., "Superconductivity of networks. A percolation approach to the effects of disorder", *Phys. Rev. B. 27*, 1983, 1541.

ALEXANDER S., "Anomalous diffusion on percolating clusters", *Physica 140 A*, 1986, 397.

GEFEN Y., AHARONY A. and ALEXANDER S., "Fractons", *Phys. Rev. Lett. 50*, 1983, 77.

RAMMAL R. and TOULOUSE G., "Random walks on fractal structure and percolation cluster", *J. de Phys. (Paris) Lett. 44*, 1983, L 13.

Hyperscaling

HAMAID T., GUYOT A., LE MÉHAUTÉ A., CRÉPY G. and MARCELLIN G., "Experimental analysis of the scaling properties of solid polymers electrolytes upon 5 years of storage: from fractal crystallization to canniens relaxation", *J. Electrochem. Soc.*, 380, 1989, 3152.

LE MÉHAUTÉ A. and CRÉPY G., "Introduction to Transfer and Motion in Fractal Media", *Solid State Ionics 9/10*, 1983, 17.

LE MÉHAUTÉ A., "Transfer Processes in Fractal Media", *J. Stat Phys. 36*, 1984, 665.

LE MÉHAUTÉ A., "Fractal Electrode", in *The Fractal Approach to Heterogeneous Chemistry*, edited by D. Avnir, J. Wiley & Sons, New York, 1989.

LE MÉHAUTÉ A., "From Dissipative to Non Dissipative Processes in Fractal Media: The Janals", in press, *New Journal of Chemistry*.

LE MÉHAUTÉ A., "Nouvelle Structure de Puits Quantiques", French patnt 89. 16190 (07/12/89).

Part II

Applications

Chapter 5

Measure and Uncertainty

We have now reached the true objective of the book: not simply to present fractal geometry, but also to show the links we have forged between this and the irreversibility of time. The following pages will show how this irreversibility comes from the uncertainty generated by measure in a fractal space.

5.1 Distribution and measure: the mathematical microscope

5.1.1 Functions with compact support

Let $\{D\}$ be the set of complex-valued functions φ defined over a compact support in the n-dimensional space of real numbers $\mathbf{R}_n$ which have derivatives of all orders. Thus any member φ belongs to the class of functions C^∞ and is identically zero outside this support.

The conditions of belonging to C^∞ and having a compact support mean that the function must have no discontinuity in a derivative of any order at the points of the boundary of the support (e.g. the end-points of an interval); this seems so drastic a requirement that it is reasonable to ask if there are such functions, that is, if $\{D\}$ is not just the empty set. But there are in fact such functions, and a simple example is

$$\begin{aligned}\varphi = \varphi(r) &= \exp\left[-1/(1-r^2)\right] \quad \text{for} \quad |r| < 1 \\ &= 0 \qquad\qquad\qquad\quad \text{for} \quad |r| > 1\end{aligned} \tag{5.1}$$

where $r^2 = x_1{}^2 + x_2{}^2 + \ldots x_n{}^2$. An important result here is the following:

> Any continuous function on a compact support, *whether derivable or not,* can be approximated by functions of the class $\{D\}$

The possibility of the function not being derivable is itself of great importance.

The functions in this set can approximate as closely as is wished, in support and size at all orders of derivation – that is, uniformly – any continuous function. The operation of going from an arbitrary initial function to φ_η is called *smoothing* and corresponds, roughly, to rounding off angles by replacing them with small arcs of radius η – an operation called "radiusing". As we might expect, the same smoothed form φ_η can result from different initial functions; we give examples of this with the definition of what is called the Dirac δ-function.

Consider the three functions of Figure 5.1:

$$
\begin{aligned}
&[\delta_0(u, \eta)]_1 = 1/\eta \quad 0 < u < \eta \\
&[\delta_0(u, \eta)]_2 = (1/\eta)\Gamma(u)\exp(-u/\eta) \\
&[\delta_0(u, \eta)]_3 = (1/2\pi)[\eta/(\eta^2 + u^2)]
\end{aligned}
\tag{5.2}
$$

All three have these properties:

1. Their integrals $= 1$: $\int [\delta_0(u, \eta)]_i \, du = 1$
2. As η tends to zero each tends zero if $u \neq 0$ and to infinity if $u = 0$:

$$
\lim_{\eta \to 0} [\delta_0(u, \eta)]_i = 0, \quad u \neq 0
$$

$$
= \infty, \quad u = 0
\tag{5.3}
$$

This limit is a quasi-function $\delta_0(u)$ called an "impulse";

$$
\delta_u = \lim_{\eta \to 0} [\delta_u(u, \eta)]_i, \quad \int \delta_0 \, du = 1
\tag{5.4}
$$

$\delta_0(u)$ is the Dirac δ-function, also called a *generalised function,* of great importance in the theory of distributions. Whilst it is obviously not a function in the usual sense of the term it is still some kind of function, since it can be approximated, to some order depending on η, by a true function $\delta_u(u, \eta)$. However, a true function, as we understand the term, is in general derivable almost everywhere: what can we say about the derivative of δ_u, and what about integration?

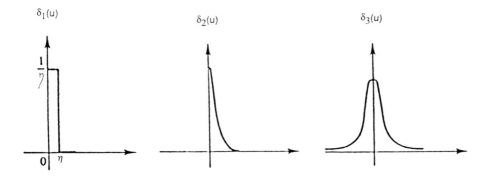

Figure 5.1 Three approximations to the Dirac function.

The approximations $[\delta_0(u, \eta)]_i$ to δ_0 are all differentiable with respect to u provided $\eta \neq 0$, so further functions can be defined by differentiating or integrating these and letting $\eta \to 0$. For example, there is what is called the Heaviside unit function $\gamma_0(u)$ (often written H(u)), defined as $\gamma_0(u) = 1$, $u > 0$, $= 0$, $u < 0$: formally, $\gamma_0(u) = \int_{-\infty}^{u} \delta(t)\, dt$, so $\delta_0(u) = \mathscr{D}_u^{1}[\gamma_0(u)]$.

So, provided that the operations make sense, the quasi-function $\delta_0(u)$ appears here as the derivative of another quasi-function $\gamma_0(u)$. These quasi-functions are called *distributions*, whose physical significance we shall be explaining later. It will be clear already that one of their fundamental characteristics is that they extend our traditional concept of a function. One can say, of course, that the Heaviside function $\gamma_0(u)$ is really quite an ordinary function: its only peculiarity is a discontinuity at $u = 0$.

It follows from the above that a singularity of the type of δ_u at the point $u = a$ can be represented by $\delta_0(u - a)$, which we can also write as $\delta_a(u)$; and a singularity of amplitude k at $u = 0$ – meaning that the integral has the value k – by $k\delta_0(u)$.

Questions naturally arise concerning the way these singularities should be used in physical applications, and here we have problems similar to those raised, for example, by the concept of measure of a non-connected fractal set. How can we comprehend an object such as the Dirac function that has no physical extension? To achieve this we use a measurement instrument called a *test function*, with which, in effect, we can give a thickness to the singularity and can test the support of a quasi-function even if this is of measure zero. The theory of distributions, with all its physical relevance, is based on the duality between the singularity and the instrument represented by the test function.

5.1.2 *Distributions*

In practical terms the theory of distributions rests on the duality between a vector space of test functions φ belonging to the set $\{D\}$ of § 5.1.1 and a corresponding vector space $\{D'\}$ of distributions. If T is a distribution, a fundamental quantity, written $\langle T, \varphi \rangle$ is a certain linear form that is continuous over $\{D\}$ and, what is important for the engineer, provides a measure of that set. We define the mapping

$$\varphi(u) \to \langle \delta(u), \varphi(u) \rangle = \lim_{\eta \to 0} \int [\delta_0(u, \eta) \cdot \varphi(u)]_i \, du \tag{5.5}$$

i.e.

$$\langle \delta_0(u), \varphi(u) \rangle = \varphi(0) \tag{5.6}$$

The mapping denoted by $\langle \cdot \rangle$ thus gives some measure of the test function $\varphi(u)$, in this case its value for $u = 0$. It is indeed a measure, because it is additive over the support – $\varphi(a \cup b) \to \varphi(a) + \varphi(b)$ – and is not identically infinite. Clearly, the value at $u = a$ is picked out by

$$\begin{aligned}\varphi(u) \to \langle \delta_0(u - a), \varphi(u) \rangle &= \lim \int [\delta_0(u - a) \cdot \varphi(u)]_i \, du \\ &= \varphi(a)\end{aligned} \tag{5.7}$$

The measuring instrument $\varphi(u)$ shows the state of the distribution $\delta_0(u - a) = \delta_a(u)$ at the point $u = a$. It is called a test function because it enables us to test, and here to measure, the distribution, which otherwise would be inaccessible since its support is of measure zero.

The treatment given so far has concerned a point distribution, that is, one defined in $\mathbf{R}_0$ with values in $\mathbf{R}_1$. However, the concept can be extended to spaces of higher dimensions, defining distributions in $\mathbf{R}_{n-1}$ with values in $\mathbf{R}_n$. We now consider an example.

5.3 *Spatial distributions: singularity of integral order > 1*

Suppose we have an infinitely thin dam – a Dirac barrage, say – whose intersection with the plane is a curve Γ. The normal to this constitutes a singularity of the type of the Dirac function, but the dam has non-zero extension in the direction perpendicular to the normal. We now try to formalise the concept of measure, using a certain Dirac-type function.

Let Γ be parametrised by a variable t, so that the function $\Gamma(t)$ defines the curve in $\mathbf{R}_2$; analogously to the previous treatment, let $\langle\delta_\Gamma, \varphi\rangle$ be the integral of $\varphi(t, u)$ along Γ, i.e.

$$\langle\delta_\Gamma, \varphi\rangle = \int_\Gamma \varphi(t, u)\, d\Gamma(t) \tag{5.8}$$

The measure for Γ can be obtained by means of the approximation to the Dirac function:

$$\langle\delta_\Gamma, \varphi\rangle = \lim_{\eta \to 0} \int_\Gamma \delta_\Gamma(u, \eta, t)\varphi(t, u)\, d\Gamma(t) \tag{5.9}$$

If we take for $\delta_\Gamma(u, \eta, t)$ the simplest form, a rectangle normal to Γ with sides η, $1/\eta$, with u the variable in the direction of the normal to the curve which is itself parametrised by t, we can find a number $\langle\delta_\Gamma, \varphi\rangle$ which is a measure of the test function φ on Γ. Then defining the concept of a tile in $\mathbf{R}_n$ by means of the test function $\varphi(t, u) = 1$ we have

$$\langle\delta_\Gamma, 1\rangle = \lim \int_\Gamma \delta_\Gamma(u, \eta, t)\, d\Gamma = \lambda \tag{5.10}$$

where λ is the length of Γ in the ordinary sense of the term.

Provided that all the necessary convergence conditions are satisfied (see the bibliography on this point) there is no difficulty in generalising this analysis, writing

$$\langle\delta_\Sigma, \varphi\rangle = \int_\Sigma \varphi\, d\Sigma, \quad \langle\delta_v, \varphi\rangle = \int_v \varphi\, dv \tag{5.11}$$

for a surface Σ and a volume v respectively.

Here the distributions δ_Γ, δ_Σ, δ_v define singularities that are not simply localised at points but have certain spatial extensions. However, the generalisation requires a normal to the path or field of integration to be defined, so that the u-axis can be defined, and this poses a problem when the path is not rectifiable: this is the case in fractal geometries, where, however, the parameter can always be defined.

The quantity $\langle\delta_\Gamma, \varphi\rangle = \int_\Sigma \varphi\, d\Gamma$ defines a physical magnitude of great importance, usually called the *circulation* of φ along the curve Γ; to investigate the nature of this, consider, for example in $\mathbf{R}_2$, a ball B of which Γ is the closed

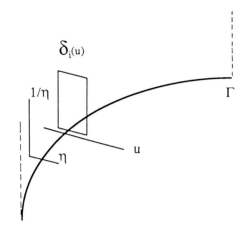

Figure 5.2 Measuring the length of the "Dirac barrage". The variable u parametrises the normal. Sweeping the rectangle $[\eta \times 1/\eta]$ over Γ gives the curve a thickness η.

boundary, $\Gamma = \partial B$. If we open Γ at the point 0 then $\langle \delta_{\Gamma - \{a\}}, \varphi \rangle = -\langle \delta_0, \varphi \rangle + K = \varphi(0) + K$, where K is a constant whose value, which can be 0, depends only on the geometry of the ball: $K = \langle \delta_\Gamma, \varphi \rangle$, which can be expressed in terms of the surface of the ball, $\int_B \beta dA = \int_{\partial B} \varphi d\Gamma - \varphi(0)$. It can be shown that there is a distribution β associated with φ which is a certain "flux" across the ball, and it is known, by Stokes Theorem, that the flux is irrotational in $\mathbf{R}_3$ if φ is the gradient of a vector. The question arises, can this treatment be extended to the case of a fractal support, because then there is no normal; we consider this at the end of this chapter.

5.1.4 *Generalisation and choice of the test function*

So far we have associated with the approximation to the Dirac function, as $\eta \to 0$, a number which is the integral of the product of two functions; generalisation of the procedure suggests that this number could be defined in other ways, for example by $\langle T_1, \varphi \rangle = \varphi(0)$, $\langle T_2, \varphi \rangle = \int \varphi(u) \, du$, $\langle T_3, \varphi \rangle = (\mathcal{D}_u^1 [\varphi(u)])_0$. We shall limit our study to distributions $T \in \{\mathbf{D}'(\mathbf{R}_n)\}$, the set of continuous linear "scalar" functionals in $\{\mathbf{D}(\mathbf{R}_n)\}$ defined by means of an integral, usually of a summable function. In other words, the distribution T associates an object, which for simplicity can be a scalar, a, with a certain function $\varphi(x)$, so that $\varphi(u) \to \langle T(u), \varphi(u) \rangle = a$.

The question arises of the meaning of the symbolism $\langle \cdot \rangle$. It is well known that the simplest number that can be associated with any function $\varphi(u)$ is the integral over its support – which is the reason for the treatment we have given. Provided that all the necessary mathematical conditions (see again the bibliography) are satisfied this can be generalised without too much difficulty and a distribution can be associated with any function by means of the integral of its product with a summable function. For example, if f is locally summable, that is, summable over any compact set, a distribution T_f can be associated with f, where

$$\langle T_f, \varphi \rangle = \int f(r)\varphi(r)\,dr = a \qquad (5.12)$$
$$\text{supp} \cdot \varphi$$

The process gives a number a which will be the same for all functions that are equal to f almost everywhere, so a single distribution T_f can be associated with a whole set of functions; and conversely it can be shown that if two locally summable functions define the same distribution they are equal almost everywhere. This suggests that the theory of distributions can be regarded as a generalisation of the theory of functions; but, like most generalisations, it tends to gloss over differences and put into the same class objects that are not in fact identical.

Our aim is to show how this fuzziness can be of especial use in developing the physics of fractal objects. In particular, we shall see that techniques based on distributions enable us to deal with the problem of the non-derivable character of a fractal interface, since the derivative of a distribution can be defined even when the number of discontinuities is so great as to prohibit the operation of elementary function theory.

In a study of the extension of the concept of derivation to a distribution it is of course possible to introduce a large number of test functions, but it is convenient to limit the choice to uniformly convergent functions. Uniform convergence implies convergence not only of the function itself, $\lim \varphi_m = \varphi$ as $m \to \infty$, but also of its derivatives of all orders:

$$\lim_{m \to \infty} \mathscr{D}^k(\varphi_m - \varphi) = 0, \quad \text{where} \quad \mathscr{D}^k = \frac{\partial^{k1 + k2 + \ldots Kn}}{\partial^{k1}\partial^{k2}\cdots\partial^{kn}} \qquad (5.13)$$

This assumption of convergence is particularly important, because it enables us to regard the number *a* as the limit to which a sequence converges; we can write

$$\lim_{\eta \to 0} \langle T(x), \varphi_\eta(x) \rangle = \langle T(x), \varphi(x) \rangle = a \tag{5.14}$$

and any continuous function on a compact support can be approximated by a test function to an accuracy that increases as η decreases: we have created a mathematical microscope. We must now investigate why it is a valuable instrument to use in connection with the non-derivable geometries that are typical of fractals.

5.1.5 *Derivatives of distributions*

Consider the distribution $\langle T_f, \varphi \rangle$ in the form

$$\langle \mathscr{D}_u^{\,1}[f(u)], \varphi(u) \rangle = \int_{R_n} \mathscr{D}_u^{\,1}[f(u)] \cdot \varphi(u)\, du \tag{5.15}$$

where $\mathscr{D}_u^{\,1}$ is the first order derivative with respect to u. Integration by parts gives

$$= [\varphi(u)f(u)]_0^{\infty} - \int_{R_n} \mathscr{D}_u^{\,1}[\varphi(u)] \cdot f(u)\, du \tag{5.16}$$

and the integrated term vanishes since $\varphi(u)$ is identically zero outside a bounded set; so we have

$$\langle \mathscr{D}_u^{\,1}[T(u)], \varphi(u) \rangle = -\langle T(u), \mathscr{D}_u^{\,1}[\varphi(u)] \rangle \tag{5.17}$$

The process can be repeated, giving the general result

every distribution T has derivatives of all orders

i.e.

$$\langle \mathscr{D}_u^{\,k}[T(u)], \varphi(u) \rangle = (-1)^k \langle T(u), \mathscr{D}_u^{\,k}[\varphi(u)] \rangle \tag{5.18}$$

This result becomes very important when the distribution has singularities, and therefore cannot be differentiated in the usual sense of the term: but a derivative can be defined in this way because the test function has derivatives of all orders. It will prove valuable in connection with, for example, charge or flux distributions having non-differentiable supports, as is the case with fractal interfaces.

It may seem that some kind of conjuring trick has been performed here, and that we have got something for nothing; later, when we come to discuss the concept of entropy, we shall see that the apparent gain has been made at the cost of a loss of local information.

The vectorial nature of the space $\{D,\}$ allows the development to be extended to the product of a distribution by an indefinitely derivable function κ, as follows:

$$\langle \kappa(T(u)), \varphi(u) \rangle = \langle T(u), \kappa(\varphi(u)) \rangle \tag{5.19}$$

This result provides the definition of the translation of the δ-function: it is most easily understood as the distribution shifted on its support by an amount a: if τ_a is the shift operator, $\tau_a[\varphi(u)] = \varphi(u-a)$ and

$$\langle \tau_a[\delta_0], \varphi(u) \rangle = \langle \delta_0, \tau_a[\varphi(u)] \rangle = \langle \delta_0, \varphi(u-a) \rangle = \varphi(a) \tag{5.20}$$

$$\langle \delta_0, \varphi(u-a) \rangle = \langle \delta_a, \varphi(u) \rangle = \varphi(a) \tag{5.21}$$

5.1.6 *Products of distributions; convolution*

The convolution operator appears naturally when we consider approximations to a fractal interface; and the concept of convolution is the key to memory in physics – that is, to the understanding of physical systems whose behaviour at a given instant depends on their previous behaviour up to that instant – they have "memory".

Let S, T be two distributions and f, g two derivable functions, all defined for the real numbers. Then there is a unique distribution W, written S ∘ T, such that

$$\langle W, \varphi(\tau, \chi) \rangle = \langle W, f(\tau) \cdot g(\chi) \rangle = \langle S(\tau), f(\tau) \rangle \cdot \langle T(\chi), g(\chi) \rangle \tag{5.22}$$

This is the *tensor product* of T and S; its value for the test function φ is

$$\langle W(\tau, \chi), \varphi(\tau, \chi) \rangle = \langle S(\tau), \langle T(\chi), \varphi(\tau, \chi) \rangle \rangle = \langle T(\chi), \langle S(\tau), \varphi(\tau, \chi) \rangle \rangle \tag{5.23}$$

It can be shown that supp $W = $ supp $T \cdot $ supp $\cdot S$. The importance of the product of distributions is that it enables us to compute the value of a quantity that has two distinct compact supports, one of which can if necessary serve as a parameter for the other. Because of this possibility of exchange between variable and parameter the product, by definition, takes account of correlations. This is expressed in a particular product called the *convolution product*, denoted by the symbol "∗". If τ, χ are variables in $\mathbf{R}_n$ we define

$$\langle S * T, \varphi \rangle = \langle S(\tau), \langle T(\chi), \varphi(\tau + \chi) \rangle \rangle$$

$$= \langle T(\chi), \langle S(\tau), \varphi(\tau + \chi) \rangle \rangle \tag{5.24}$$

If $\tau + \chi = a$, constant, so that $\chi = a - \tau$, then $\langle S(\tau), \langle T(a - \tau), \varphi(a) \rangle \rangle$ is constant. This is a scalar related to the fixed point a of the convolution. The result means that if we wish to know the value of this constant we must know all the information carried by the parameter τ.

Figure 5.3 shows how the convolution sets up correlations and, by definition, envisages them all.

Various conditions related to compactness have to be fulfilled, but we shall not go into this here, and simply assume that the convolution always exists; this is certainly so if the distributions are defined by summable functions:

$$\langle f * g, \varphi \rangle = \iint f(\tau) g(\chi) \varphi(\tau + \chi) \, d\chi \, d\tau$$

$$\langle f * g, \varphi \rangle = \iint f(\tau) g(\tau - t) \varphi(t) \, dt \, d\tau \tag{5.25}$$

$$\langle f * g, \varphi \rangle = \langle \int f(\tau) g(\tau - t) \, d\tau, \varphi(\tau) \rangle$$

$$h(t) = f * g = \int_0^t f(\tau) g(\tau - t) \, d\tau = \int_0^t g(\tau) f(\tau - t) \, d\tau \tag{5.26}$$

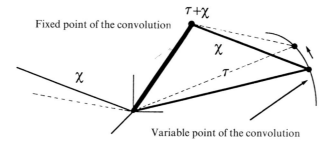

Figure 5.3 Representation of the correlations on which the convolution is based: Equation (5.24).

To help understand the physical significance of the convolution we consider the following physico-chemical example. A chemical driving force $\Delta X(t)$ – for example, a raised temperature or a reduced pressure – is applied at time $t = 0$, in a suitable reactor, to a wire covered with a sublimable oxide. We want to know the rate $J(t)$ at which the material is being extracted at time t. The function representing the driving force can be broken down into a large number of small steps, as in Figure 5.4.

$J(t)$ is related to the length L of the wire by $J(t) = L\varphi(t)$, where $\varphi(t)$ is the surface flux density function which is linearly related to $\Delta X(t)$; all the parameters are constant.

If L is constant, an initial flux increment is given by $\Delta J(t) = \langle L \cdot \langle \varphi'(0) \cdot \gamma(t) \rangle \rangle$, where $\gamma(t)$ is the Heaviside unit step function defined in § 5.1, $\varphi(0)$ is the height of the step and $\varphi'(0)$ is the ordinary derivative of $\varphi(t)$ at $t = 0$; whilst if L varies in such a way that its length $L(t)$ is the transfer function between the flux density and the total flux then $\Delta J(t) = \langle L(t) \cdot \langle \varphi'(0) \cdot \gamma(t) \rangle \rangle = \langle L(t), \langle \varphi'(t), \gamma(0) \rangle \rangle = \langle L(t), \langle \varphi(0) \cdot \delta_t \rangle \rangle$. With this assumption the length, which is a geometrical parameter, constitutes the basic transfer function which explains the kinetics of this physico-chemical system. Starting from this initial state, we apply a small step $\Delta \varphi$ of driving force $\Delta \varphi(\Delta \tau) \cdot \gamma(\Delta \tau) = k \cdot \Delta X(\Delta \tau)$ at time $t - \Delta \tau$. The response of the system at time t is $\Delta J(t) = L(t - \Delta \tau) \cdot \Delta \varphi(\Delta \tau) \cdot \gamma(\Delta \tau)$, since by the linearity assumption $L(t)$ responds in the same way at every instant. The same argument applies for $2\Delta \tau$, $3\Delta \tau$, ... $n\Delta \tau$, and by addition we have

$$J(t) = \sum_n \Delta J(t) = L(t)\varphi(0)\gamma(0) + L(t - \Delta \tau)\Delta \varphi(\Delta \tau)\gamma(\Delta \tau)$$

$$+ L(t - 2\Delta \tau)\Delta \varphi(2\Delta \tau)\gamma(\Delta \tau) + \ldots\ldots\ldots\ldots\ldots + L(n\Delta z)\Delta \varphi(t)\gamma(t) \quad (5.27)$$

Thus for $t > 0$

$$J(t) = \sum_n \Delta J(t) = L(t)\varphi(0) + L(t - \Delta \tau)\Delta \varphi(\Delta \tau) + L(t - 2\Delta \tau)\Delta \varphi(2\Delta \tau)$$

$$+ \ldots\ldots\ldots + L(0)\Delta \varphi(t) \quad (5.28)$$

If we increase the number of steps indefinitely, decreasing their size correspondingly, $\Delta \varphi$ becomes $(\partial \varphi/\partial \tau)\,d\tau$ and (5.28) becomes

$$J(t) = E(0)L(t) + \int_0^t L(t - \tau)\varphi'(\tau)\,d\tau \quad (5.29)$$

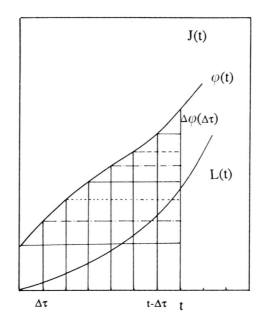

Figure 5.4 Illustration of (5.28). $\varphi(t)$ is broken into steps $\Delta\Phi(\Delta\tau)$ before being multiplied by $L(t-\tau)$. $L(t)$ is assumed independent of the time origin, and to have a linear response.

This can be written

$$J(t) = E(0)L(t) + \int_0^t L'(t-\tau)\varphi(\tau)\,d\tau = E(0)L(t) + \mathscr{D}_t^1\left[\int_0^t L(t-\tau)\varphi(\tau)\,d\tau\right]$$

(5.30)

where as usual $\mathscr{D}^1$ is the first-order derivative. The above formulae constitute a theorem of Vaschy which can be written symbolically as

$$J(t) = L'(t) * \varphi(t) = L(t) * \varphi'(t)\,(L' = dL/dt\ \text{etc})$$

(5.31)

This is the very simple basis of the TEISI and TEISA models, mentioned in section 2.3.

Notice that whilst $\varphi(t)$ need not be continuous, it must be bounded; and $L(t)$ has to be derivable if and only if the input signal – that is, either $J(t)$ or $\Delta X(t)$, depending on the physical arrangements – is not derivable. Thus if we wish to

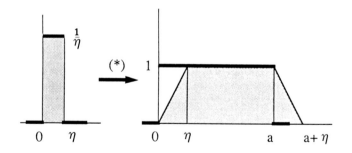

Figure 5.5 $\alpha_\eta * T$, an example of smoothing. Examples of smoothing by Fourier transforms are given in Angot 1967.

know the response of the system at time t we must know its behaviour up to this time – its past history, in fact; the convolution explains on the one hand the existence of this memory and the accompanying smoothing, and on the other hand the irreversibility that is associated with this. It assumes also the existence of a stable transfer function, that is, a linear behaviour. One might say that the convolution makes of the mathematician something of an historian: to understand the present he must know of the past – and he cannot reverse the flow of time.

5.1.7 *Smoothing by convolution: dealing with uncertainty*

Consider the very simple example of convolution with the Dirac function:

$$\langle \delta * T, \varphi \rangle = \langle T, \langle \delta, \varphi \rangle \rangle = \langle T, \varphi \rangle \tag{5.32}$$

This shows that $\delta * T = T \cdot \delta$ is the neutral element for the convolution operation. To understand in what way convolution effects a smoothing we show, using a simple example, how it works with an approximation to the Dirac function.

Let α_η be the distribution whose value if $1/\eta$ on $[0, \eta]$ and T the distribution with value 1 on $[0, a]$. Figure 5.5 shows these and that the convolution $\alpha_\eta * T$ is a continuous function that is differentiable almost everywhere.

Clearly, the convolution operation has resulted in some loss of information since it has "smoothed out" irregularities. In particular, our knowledge must be limited to the surface properties of the ball: for a given approximation "η" we are in complete ignorance of the properties within the approximating ball, and in

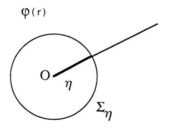

$\varphi(r)$

Figure 5.6 The value of the test function at O can be found by evaluating it over the surface Σ_η of the ball of radius η and letting $\eta \to 0$.

fact these have been made homogeneous. This is the fundamental explanation of equation 2.36 which defines ergodicity in a fractal medium: it is fruitless to look for the behaviour on a microscopic scale.

We use approximations to the Dirac function again to help understand the implications of the convolution. Suppose we have an approximating ball with centre O and radius η; let Σ_η be its surface and $\varphi(r)$ a radial function defined in its interior, as in Figure 5.6. Let S_η be the distribution defined by

$$\langle S_\eta, \varphi(r) \rangle = (1/4\pi\eta^2) \int_{\Sigma_\eta} \varphi \, d\Sigma = (1/4\pi\eta^2) \langle \delta_\Sigma, \varphi \rangle \tag{5.33}$$

This is the mean value of φ over the surface of the ball; integration by parts gives

$$\langle S_\eta, \varphi(r) \rangle = (1/4\pi\eta^2) \left[(\varphi(r) \cdot \Sigma_\eta) - \int_{\Sigma_\eta} \Sigma_\eta \, d\varphi \right]$$

$$\langle S_\eta, \varphi(r) \rangle = (1/4\pi\eta^2) \left[(4\pi\eta^2)\varphi(\eta) - \int_{\Sigma_\eta} \Sigma_\eta \, d\varphi \right] \tag{5.34}$$

If $\eta \to 0$ the integral over Σ_n tends to zero and we have

$$\lim_{\eta \to 0} \langle S_\eta, \varphi(r) \rangle = \varphi(0) \tag{5.35}$$

and hence

$$\lim_{\eta \to 0} S_\eta = \delta_0 \tag{5.36}$$

Thus the limit of the approximation to $\varphi(r)$ by S_η as $\eta \to 0$ is the measure of $\varphi(r)$ *at the centre of the ball*; so the distribution δ, localised at a singularity, can be regarded as the limit of a distribution S_η defined over the surface of a ball centred at that singularity. In one very particular case, the Laplacian case, the value of the test function $\varphi(r)$ at the surface of the ball is equal to its Euclidean mean over the interior.

The choice of approximations α_η to the Dirac function is unlimited, but they should have these properties:

1. positive: $\alpha_\eta(x) \geqslant 0$
2. radial: $\alpha_\eta(x) = f(r)$ where $r = |x|$
3. derivable to all orders
4. $\operatorname{supp} \alpha_\eta(x) = B_h(0)$, the ball of centre 0 and radius η
5. of unit value for all η, i.e.

$$\langle \alpha_\eta, \mathbf{1} \rangle = 1 \text{ (where } \mathbf{1} \text{ is the unit } \mathbf{R}_n) \tag{5.37}$$

Here the ball is treated as representing the arithmetical unit.

Given these properties, α_η can be used to smooth out a discontinuity in the support of a distribution. For example, let

$$\alpha_\eta = 3/4\pi\eta^3 = 1/\tau_\eta \quad \text{for all x within the ball}$$
$$= 0 \text{ elsewhere} \tag{5.38}$$

This definition meets all the above requirements.

The convolution product is $\alpha_\eta * \varphi(r) = (1/\tau_n)\int_{B_\eta} \varphi(r)\, dr$, from which we can compute the mean of the distribution $\varphi(r)$: it is in fact its value, smoothed by the radial distribution $\alpha_\eta(r)$, at a point "0" on the ball $B_\eta(0)$. As $\eta \to 0$ the convolution product tends to the measure of $\varphi(r)$ at the point 0:

$$\lim_{\eta \to 0} [\alpha_\eta(r) * \varphi(r)] = \lim_{\eta \to 0} \left[1/\tau_\eta \int_{B_\eta} \varphi(r)\, dr \right] = \delta_0 * \varphi(r) = \varphi(0) \tag{5.39}$$

This shows how the property can be made use of when the singularity is distributed over a fractal support: to understand the behaviour at the fractal singularity it is sufficient, after having smoothed the support, to find the behaviour at the limit of the approximation to the fractal. In this context, consider this second example: let φ be the flux across an interface Γ and S_η the

Dirac approximator previously defined; with all the limits as $\eta \to 0$:

$$\lim \langle S_\eta * \delta_\Gamma, \varphi \rangle = \lim \langle S_\eta, \langle \delta_\Gamma, \varphi \rangle \rangle = \lim \langle S_\eta, \varphi \langle \delta_\Gamma, 1 \rangle \rangle \qquad (5.40)$$

$$= \langle \delta_0, \varphi \cdot \lambda_\Gamma \rangle = \langle \delta_0, J \rangle \qquad (5.41)$$

where λ is the length of the interface Γ and J is the total flux transferred across it. Now

$$<\delta_\Gamma \langle \lim S_\eta, \varphi \rangle = \int_\Gamma (\lim (1/4\pi\eta^2) \int_\Sigma \varphi \, d\Sigma) \, d\Gamma \qquad (5.42)$$

which means that this total flux is the integral along the interface Γ of the mean flux density (the test function φ) over the surface Σ_η of the approximating ball, in the limit as $\eta \to 0$.

In this calculation the convolution with φ has explicitly regularised the interface by means of the approximator. Thus if we were studying a distribution T which we expected to have very many discontinuities – as would be the case for a charge distribution over a fractal support – we should need to form the convolution of this with a ball of radius η, for example $\delta(t, \eta)$, and take the limit as $\eta \to 0$.

The convolution operation is smoothing since in a sense it eliminates the irregularities in centering the approximating ball on the fractal. The reader will see the clear analogy here between convolution with the Dirac function and the measure of a fractal support; we shall develop this analogy further, ultimately to give a physical basis, in terms of measure, for the assumptions concerning the convolution of section 2.3.

5.1.8 *Experiments and distributions: an example*

As we have just seen, the importance of the concept of a distribution is due to the ability it gives us to deal with singularities, in a sense by smoothing them out. This ability is not merely formal but reflects a very fruitful analogy between physics and the mathematics underlying the theory of distributions, an analogy which in turn rests on the identity of the two uses of the test function $\varphi(r)$: for *smoothing* and, as a microscope, for *detection*.

Thus we have:

Test function φ		
Distribution T	$\rightarrow$	Linear scalar form a = $\langle T, \varphi \rangle$

Experimental method $\downarrow$ $\uparrow$

Experimental tool φ_η		
State function on support T	$\rightarrow$	Measure, approximation η

We can illustrate this analogy by means of an example due to Pinchard. The problem is to find the electric potential difference ΔU between the ends a, b of an infinitely thin conducting wire Γ_{ab}. We take as test function the distance smoother φ_η defined by

$$\varphi_\eta = \mathrm{Sm} \cdot [t \cdot \delta_\Gamma] = t \cdot \delta_\Gamma * \alpha_\eta \qquad (5.43)$$

where t is the arc length measured along Γ and α_η is the radial function (5.38) defined over the ball B_η. Consider the intersection of B_η with Γ and let Δl_η be the length of the part of Γ within B_η: we have

$$\Delta l_\eta = \Gamma_{ab} \cap B_\eta \qquad (5.44)$$

then

$$\varphi_\eta(u) = (1/v_\eta) \int_{\Delta l_\eta} t \, dt, \quad v_\eta = 4\pi\eta^3/3 \qquad (5.45)$$

This last expression is simply the mean value of the measure over the section of arc distributed in mean over the volume of the ball; approximately

$$\varphi_\eta(t) \approx (\Delta l_\eta/v_\eta)t_\eta \qquad (5.46)$$

Now suppose an electric field E is applied to φ_η; we get, as an approximation of order η to the electromotive force at the ends of the arc.

$$U_\eta = \langle E, \varphi_\eta \rangle \sim \int_\Gamma (E * \alpha_\eta)t \, dt \qquad (5.47)$$

in which the term E α_η is the spherical mean, weighted by α_η, of the distribution of field E computed over the ball B_η centered on the arc Γ_{ab}. If E is a continuous function and if v_η is the measure of B_η then

$$E * \alpha_\eta = (1/v_\eta) \int_{B_\eta} (E * \alpha'_\eta) \, dv \tag{5.48}$$

that is, U_η represents the mean integral of the field E over a tube of radius η surrounding Γ.

This study recalls a valuable development described by Pinchard in 1977, in his course on generalised functions in electromagnetism, which can be summarised as follows. If a source (a field) is represented by a distribution E (a generalised function), an experiment will be represented by a test function φ_η which smooths the support of this distribution by means of a suitable approximation to the Dirac function. The result of the experiment will depend on the smoothing.

Experimental tool φ_η		
Intensive quantity E(t), state	$\rightarrow$	Approximate result U_η

Thus the result is always approximate. However, the interaction of a distributed field with an experimental device can always be interpreted as the response U_η of a physical system E(t) observed imperfectly by means of a test function Φ_η. In particular, the experimental tool is a system whose geometry is fixed, and it is this geometry that produces the uncertainty in the result. If we concern ourselves not with the state of the system but with changes to this "excitations", what we shall get is the response to an enquiry $\Delta X(t)$ rather than the result of a static measurement. There is thus a duality between the state of the system and the means by which this is evaluated;

Excitation (input) $\Delta X(t)$		
System φ_η	$\rightarrow$	Response U_η

This is the basis of the experimental procedure, whether on the macroscopic or microscopic scale. The analysis just given is well known in particle physics; fractal geometry leads naturally to its application at the microscopic level.

ANGOT A., *Compléments de Mathématiques à l'usage des Ingénieurs de l'Electrotechnique et des Télécommunications*, Masson, Paris, 1967.

ARSAC J., *Transformations de Fourier et Théorie des Distributions*, Dunod, Paris, 1961.

EUVRATD D., "Distributions", cours de l'ENSTA, 1977.

GABOR D., "Theory of the communication", *J. SEE 93*, 1946, 429.

GELFAND I.M., CHILOV G.E. and VILENKIN N.Y., *Les Distributions*, Dunod, Paris, 1967.

RODIER F., *Distribution et Transformation de Fourier*, McGraw-Hill, Paris, 1978.

SCHWARTZ L., *Méthodes Mathématiques pour les Sciences Physiques*, Hermann, Paris, 1965.

SCHWARTZ L., *Théorie des Distributions*, Hermann, Paris, 1966.

ZEMANIAN A.H., *Distribution Theory and Transform Analysis*, McGraw-Hill, New York, 1965.

5.2 Dimension of the Weierstrasse distribution

This has been discussed already in Chapter 3 (§ 3.3.2, equation 3.15). Weierstrasse invented the distribution in order to show that functions can exist that are continuous but nowhere derivable:

$$W(x) = \sum_{n=0}^{\infty} v^{-nH} \cos(v^n x)\, (v > 1, H > 0) \tag{5.49}$$

This is shown (approximately) in Figure 5.7.

This is a Fourier series in which the successive terms have frequencies that increase as powers of v (> 1) and are weighted not by constants but by an exponentially decreasing function.

Thus $W(x)$ is a sum of sinusoids that become ever sharper as n increases, and their amplitudes ever smaller. The spectrum is discrete, with critical frequency $f_n = v^n$. The power spectrum $G(v^n)$ is proportional to the square of the amplitude, i.e. $G(v^n) \sim v^{-2nH}$; so the graph of G against $\log v^n$ is a straight line of slope 2H and G belongs to the class of generalised functions known as $1/f$ noise, in this case $1/f^{2H}$.

We can use the boxes method of § 2.2.1 to find the dimension of the graph of $W(x)$; write

$$W(x) = \left(\sum_{n=0}^{k-1} + \sum_{k}^{\infty} \right) v^{-nH} \cos(v^n x) = W_0 + W_k \tag{5.50}$$

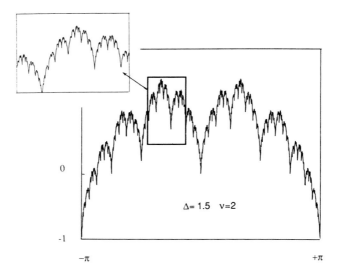

Figure 5.7 Weierstrasse distribution for $v = 2$, $\Delta = 1.5$, showing the non-derivability. The enlargement can be repeated indefinitely.

The oscillations in W_k are of higher frequency and smaller amplitude than those in W_0; taking the first term as an approximation, in unit time there are v^k oscillations of amplitude $v-kH$ and width v^{-k}. If we cover the graph with squares of side $v-k$ we find

$$N(k) = v^k \cdot v^{-k} \cdot v^{-kH}/v^{-k} \cdot v^{-k} = v^{(2-H)k} \tag{5.51}$$

By definition, the fractal dimension is

$$\Delta = \lim_{k \to \infty} [\log N(k)/\log vk] = 2 - H \tag{5.52}$$

This can be rewritten as $H = 2 - \Delta$, so H is the co-dimension of the graph, that is, the dimension of the content of its environment.

5.3 Local Hölderian properties: geometrical meaning of derivatives of non-integral order

As was stated in § 3.3.2, the derivative of order α of the Weierstrasse function

W(x) exists everywhere if $H > \alpha$, and nowhere if $H < \alpha$; so it is differentiable everywhere in the ordinary sense ($\alpha = 1$) if $H > 1$ and nowhere if $0 < H < 1$. The graph is particularly interesting if H is in this open interval, for then in the neighbourhood of any x_0 it lies between two parabolic-type curves (they are strictly parabolas if $H = 1/2$) with vertex x_0. This, called the Hölderian property, is illustrated by Figure 5.8.

The approximation by parabolic-type curves generalises the concept of the tangent, and thus that of speed: if the curve is derivable at a point x_0 then in that neighbourhood it lies within a cone of arbitrarily small angle. The general property is related to those of fractional derivation and also to the theory of generalised functions. For example, the integral

$$_{-\infty}\mathscr{D}_t^{\alpha-1}[W(t)] = \int_{-\infty}^{t} w(z)(t-z)^{-\alpha}\,dz, \tag{5.53}$$

which smooths W(t) is the convolution product of W with $X_{-\alpha}(t) = t^{-\alpha}$, that is a smoothing of the graph by means of balls of content $t^{-\alpha}$. If $0 < \alpha < 1$, the function defined by the convolution $W * X_{-\alpha}$ is such that the derivative of W of order α exists.

Thus fractional derivation arises from convolution with a function $X_{-\alpha}(t)$ which is a (non-integral) power of t. We are interested in the relation non-integral derivation and the fractal properties of the graph in the plane. Mathematically, this is provided by the local Hölderian properties. Suppose that for all t

$$0 < C1 < |W(t_0) - W(t)|/|t_0 - t|^{\alpha} < C2 < \infty \tag{5.54}$$

This defines a generalised derivative which can be interpreted as a generalised speed along the curve, replacing the traditional concept of speed. Thus there is a close relation between the three concepts, as follows:

Non-integral derivation $\leftrightarrow$ Hölderian property $\rightarrow$ fractal dimension

5.4 The NOISE transformation

As was shown in Chapter 2, the space-time dispersion $(l(t), \eta(p))$ that underlies uncertainty originates in the parametrisation of the geometrical interface. Provided that this parametrisation depends on the metric of the space – in the

TEISI model internal via the dimension, in the TEISA model external via the co-dimension – the uncertainty in Laplace space and in space-time is given by the measure gauges

$$\eta(p) = \lambda_0 p^{-\beta} \rightarrow l(t) = \lambda_0 t^{\beta-1} \tag{5.55}$$

This shows the importance of the parameter λ_0, a fractional conjugate of time with respect to space: it is a speed if $\Delta = 1$ and a diffusion coefficient if $\Delta = 2$; and if $\Delta = 1/2$ then λ_0^2 is an acceleration, cf. in *Janals*. Thus λ_0 plays the role of a "zoom": as it increases, so does the uncertainty.

The models referred to are based on a constant value of λ_0 and thus on a fixed level of uncertainty. Complementary models can be developed in which any value of λ_0 is allowed; to illustrate the consequences of this we look at the response of certain distributions – Dirac, Cantor, Weierstrasse – under the effect of the uncertainty resulting from the parametrisation of the set on which the generalised function is defined.

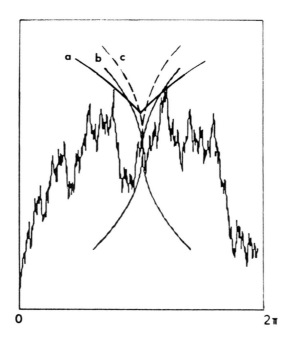

Figure 5.8 Representation of the physical meaning of fractional derivation, relation to the fractal character of the Weierstrasse distribution. a, b, c correspond to different exponents of the power law.

Consider for example a binary Cantor distribution $\varphi(t)$. Every measure of this requires a smoothing by means of a test function; if we take for this a power function $X_\alpha(t) \sim |t/a|^\alpha$ the smoothing can on the one hand be associated with a fractional integro-differentiation operation, and on the other, through the parameter a, perform an enlargement as though with a microscope. The response of the system is thus of the form J(t), where

$$J(t) = \varphi(t) * X(t) \tag{5.56}$$

The study of the uncertainty resulting from the variation of the parameter a leads to the simulation technique called NOISE (NOn-integral Integration for Simulation of Experiments), based on the application of a power-law test function to the distribution of the systems studied. The resulting convolution gives the amplitude of the response, taking into account the uncertainty in the support of the generalised function.

Figure 5.9 Support for the Cantor distribution.

5.5 Wavelet transform of a fractal object

The *wavelet transform* technique is a recent development, mainly for seismological applications; it gives a 2-dimensional representation of an originally 1-dimensional signal. The method is based on the development of an arbitrary function s(t) in terms of a base set of orthogonal functions $g_a(t) = g(t/a)$ called wavelets, derived from a single base function by translation and dilatation. The transform process is a convolution – and thus a measure – together with a smoothing: the wavelet transform $T_s(a, t)$ of s(t), with $a > 0$, is

$$\begin{aligned} T_s(a, t) = g_a(t) * s(t) &= \int g_a(t - \tau)s(\tau)\,d\tau \\ &= \int g[(t - \tau)/a]s(\tau)\,d\tau \end{aligned} \tag{5.57}$$

Like all convolutions this transform constitutes a local measuring instrument, pointed at t = τ; it is also a "zoom", the factor a acting as a magnifier. It differs from NOISE in that there is no underlying ergodic assumption: it is a mathematical technique, which can reveal the source of the fractal structure. The zoom makes the treatment less abstract.

ARNÉODO A., GRASSEAU G. and HOLSCHNEIDER M., *Phys. Rev. Lett. 612281*, 1988.

ARNÉODO A., GRASSEAU G. and HOLSCHNEIDER in *Wavelets*, (edited by J.M. Combes, A.).

FRISCH U., SULEM P. and NELKIN M., *J. Fluid. mech.87 1978, 719.*

GABOR D., "Theory of the communication", *J. SEE 93*, 1946, 429.

GROSMAN A. and MORLET J., in "Mathematics and physics, lecture on recent results" (edited by L. Streit), *World Scientific*, Singapore, 1987.

KOLMOGOROV A.N., *C.R. Acad. Sci. URSS30, 301*, 1941.

MANDELBROT B.B., *J. Fluid. Mech. 62, 331*, 1974.

PARISI G. and FRISCH U., in *Turbulence and Predictability in Geophysical Fluid Dynamics and Climate Dynamics* (edited by M. Benzi and G. Parisi), North Holland, Amsterdam, 1985, (GROSSMAN and P. TCHAMITCHIAN), Springer-Verlag, Berlin, in press.

LE MÉHAUTÉ A., "Etude de la structure de l'électrode de piles pour batterie et accumulateur", Rapport des laboratoires de Marcoussis, 1980, unpublished.

THOMSON A.H., Katz A.J. and C.E., *Adv. in Phys. 36(5)*, 1987, 625.

NIGMATULLIN R.R., *Phys. Stat. Sol. (b)*, 153, 1989, 49.

Chapter 6

Fractal morphogenesis

Many natural objects have fractal structures, their fractal character resulting from the irreversibility of the processes that generate them. Although this irreversibility is not always emphasised, it lies at the heart of fractality; in real-life conditions, if a process cannot dissipate its energy the only way in which it can continue is to destroy the metric, that is, to create a fractal metric.

In this chapter we give some simple illustrations of this fractal morphogenesis. We shall not consider the energy balance, but we must say at the outset that this is essential to the understanding of the physical phenomena that we describe.

6.1 Fractal structure of manganese dioxide

The photographs of Plate 1 are not of clouds as might be supposed, but of the manganese dioxide from a perfectly ordinary dry battery. The magnifications cover two orders of magnitude, but the individual values are not given and cannot be deduced from the photographs: this is another way of saying that the material is fractal. The ability to assume a fractal state seems to be a characteristic of heterogeneous materials; and it seems that this governs their behaviour in industrial processes.

LE MÉHAUTÉ A., "Etude de la structure de l'electrode de piles pour batterie et accumulateur", Rapport des laboratoires de Marcoussis, 1980, unpublished.
THOMSON A.H., Katz A.J. and C.E., *Adv. in Phys. 36*(5), 1987, 625.
NIGMATULLIN R.R., *Phys. Stat. Sol. (b)*, 153, 1989, 49.

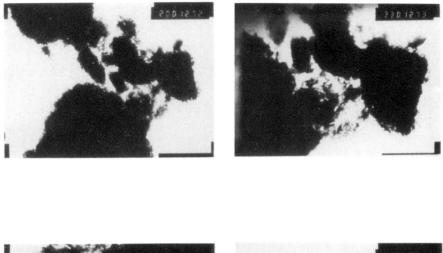

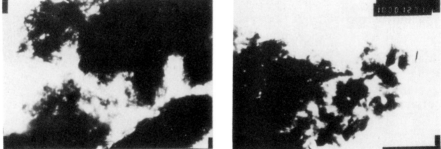

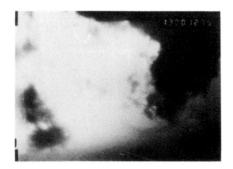

Plate 1. Fractal structure of manganese dioxide (MnO_2). Photographs at a series of magnifications simulate a zoom-in on the material (6.1). (Photographic credit: Laboratoires de Marcoussis, C.R. de la C.G.E.)

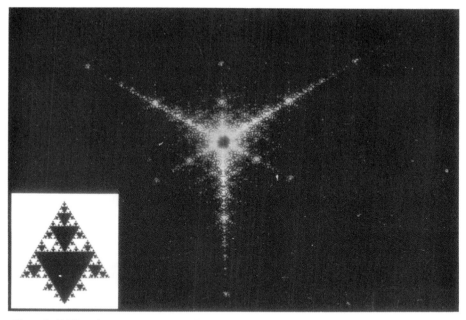

Plate 2. Diffraction pattern produced by a 2-dimensional fractal object (a Sierpinski Sieve) (6.2). (Photographic credit: Laboratoires de Marcoussis, C.R. de la C.G.E.)

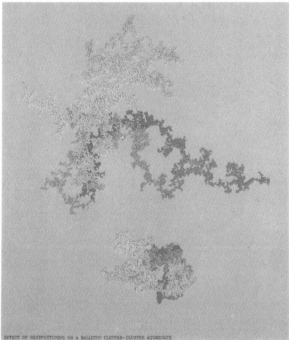

Plate 3. Clusters and aggregates generated by numerical simulation (6.3). (R. Botet and R. Jullien, Physics of Solids Group, Université de Paris Sud, Orsay).

6.2 Diffraction by a fractal object

Suppose we illuminate a fractal object whose mass and surface dimensions are Δ_M, Δ respectively (cf. Chapter 1, §1.5) by coherent light of wavelength λ. The light will be diffracted by the object and the incident energy dispersed in all directions in a manner that depends on the angle of incidence θ. The diffraction is due to a transfer of momentum characterised by a length $\eta = (4\pi/\lambda) \sin (\theta/2)$.

Figure 1.9 is an example of the images produced by such a process. It can be shown that the intensity $I(\theta)$ of the light scattered in the direction θ follows the law $I(\eta) \sim \eta^{-\gamma}$, with $\eta = \eta(\theta)$ as above. This law holds for diffraction in what is called the "Porod" region, characterised by $1/\eta$ being less than the characteristic length of the diffracting object. The exponent γ depends on the structure of the object:

- if the diffraction depends on the mass of the object, which will vary like R^{Δ_M}, where R is a characteristic dimension, then $\gamma = \Delta_M$
- if the fractal structure is defined by the surface, and therefore by Δ, the diffraction does not depend on the mass and now $\gamma = 2d - \Delta = d + \zeta$, where d is the dimension of the space in which the object is embedded (2 for a plane, 3 for ordinary space) and ζ is the co-dimension.

Plate 2 is an example of a diffraction pattern.

CHABASSIER G., HELIODORE F. and LE MÉHAUTÉ A., *J. Optics Am. Soc.*, in press.
KJEMS J.K. and SCHOFIELD P., in "Scaling Phenomena in Disordered Systems", NATO ASI, Plenum, New York, 1986.
Proceedings of the Royal Society of London, *Fractal in the Natural Sciences*, *423*, 1989, 1, edited by M. FLEISCHMANN, F.R.S., D.D. TILDESLEY and R.C. BALL, October, 1988.
SCHAEFER D.W. and KEEFER K.D., *Phys. rev. Lett. 53*, 1984, 1383 and *Mat. Res. Symp. 32*, 1984, 1.
SCHAEFER D.W., MARTIN J.E., HURD A. J. and KEEFER K.D., *Physics of Finely Divided Matter*, Springer-Verlag, New York, 1985, 31.

6.3 Growth of aggregates

The first model for this process was what is called the DLA (diffusion-limited aggregation) model, developed in 1981 by Witten and Sander as part of a research project undertaken by the Exxon company, with the aim of explaining the growth of fractal objects. It considers particles in a set E that undergo Brownian motion and on reaching the boundary of E diffuse on to a cluster G, with which they aggregate. A particle at the boundary that leaves E is replaced by another that follows in its turn, and so on; so the aggregation occurs stage by stage. The model describes a system that is not in equilibrium.

The aggregates thus formed grow preferentially by developing sharp protruberances which divide and develop at each stage. In the original model only the particles move, the clusters remaining fixed; the fractal dimension of the clusters formed by the process is approximately 1.7.

In a development of the model the clusters also are allowed to move thus introducing the possibility of clustering among themselves: this accounts very well for such physical phenomena as aggregation in colloids. The cluster-to-cluster aggregation is clearly hierarchical: similar-sized clusters aggregate to form larger clusters, and so on, and is better able than the original DLA model to explain why the objects generated are fractal. (See plate 3).

BOTET R., JULLIEN R. and KOLB M., *J. Phys. A17*, 1984, L75.

BOTET R. and JULLIEN R., "Aggregation and Fractal Aggregates", *World Scientific*, Singapore, 1987.

BOTET R. and JULLIEN R., *Ann. Phys. 13*, 1988, 153.

JULLIEN R., *Ann. télécommun. 41*, 7/8, 1986, 343.

KOLB M., BOTET R. and JULLIEN R., *Phys. Rev. Lett. 51*, 1983, 1123.

LUBENSKY T.C. and PINCUS P.A., *Phys. Today 37 (10)*, 1984, 44.

MAZEKIN P., *Phys. Rev. Lett. 51*, 1983, 1119.

MEAKIN P. and WASSERMAN, *Phys. Lett. A 103*, 1984, 337.

SANDER L.M., *Nature 322*, 1986, 789.

SANDER L.M., *Sc. Am. 256 (1)*, 1986, 94.

SMIRNOV B.M., *Sov. Phys. Usp 29*, 1986, 481.

WITTEN T. and SANDER L., *Phys. Rev. Lett. 47*, 1981, 1400 and *Phys Rev. B27*, 1983, 56.

WITTEN T. and CATES M.E., *Sciences 232*, 1986, 1607.

6.4 Hydrodynamical fingering

If a fluid of relatively low viscosity is injected into a medium of higher viscosity contained between parallel plates, finger-like instabilities are formed at the interface, branched to a greater or less extent: this is called viscous fingering, VF. If the receiving medium is visco-elastic and the difference in the viscosities is

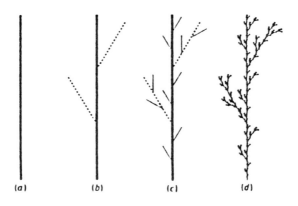

Figure 6.1 Development of a tree structure of dimension given by Equation 6.1.

great then these fingers have a fractal structure very similar to that of the aggregates generated by the DLA model described in § 6.3. This morphological similarity is due to the fact that although the two physical processes are very different, the laws governing them are of the same type. In the viscous case the rate of growth of a finger is proportional to the local pressure gradient (Darcy's Law), whilst in the diffusion case the rate of growth of the aggregate is proportional to the concentration gradient (Fick's Law).

The analogy is seen also in the dynamics of the processes: in both the aggregates and the fingers growth occurs preferentially at the extremities, to the detriment of the bays and gulfs between the protruberances.

The study of the viscous-flow phenomenon has an important application in the petroleum industry, to the flow of oil through porous rock of low permeability. A similar approach has led to a method for predicting the movement of the oil/water interface in the process of recovering oil by injection of water.

Still other physical processes show an analogy with the DLA model, for example the growth of crystals or of needles under the influence of a temperate gradient; the relevant field in this case is the electric potential. Whatever the mechanism assumed, the structures generated in the 2-dimensional case have a factal mass dimension of around 1.7, in agreement with the DLA model.

An analysis of the finger problem based on self-similarity shows that the fractal dimension can be expressed in terms of certain ratios, as follows:

$$\Delta = -(\log r_n)/(\log r_1) \tag{6.1}$$

where r_n (> 1) is the ratio of the number of fingers at stage n to the number at stage $n-1$; and r_1 (< 1) is a ratio describing the change in the lengths of the fingers between these stages (cf. Fig. 6.1). The value found for Δ for the 2-dimensional viscous case is close to 1.5.

DACCORD G. and LENORMAND R., *Nature 325*, 1987, 41.

HITTMANN J., DACCORD G. and STANLEY H.E., *Nature 314*, 1985, 141.

VAN DAMME H., "Flow and Interfacial Instabilities in Newtownian and Collodïdal Fluid" in *The Fractal Approach to Heterogeneous Chemistry*, edited by D. Avnir, J. Wiley and Sons, New York, 1989.

VAN DAMNE H. ALSAC E. and LAROCHE C., *C.R. Acad. Sci. Paris 309* (II), 1989, 11.

VAN DAMNE H., OBRECHT F., LEVITZ P., GATINEAU L. and LAROCHE C., *Nature 320*, 1986, 731.

VAN DAMNE H., LAROCHE C., GATINEAU L. and LEVITZ P., *J. Phys. 48*, 1987, 1121.

VAN DAMNE H. and LEMAIRE E., in press.

6.5 Electrical discharge fingering

As with energy dissipation by viscosity, dissipation by electrical discharge leads to the development of finger-like structures. However, the topology and the metric of these structures differ according as the discharge is produced by a negative or a positive potential difference. For a negative discharge these resemble those due to viscosity, for positive the structures seem to have the limiting topology of a phenomenon that depends on the volume of the object, which can be shown to involve the co-dimension. Many questions remain to be answered, in spite of the excellent experimental studies that have been and still are being made, in particular concerning the relation between the irreversibility of the discharge and the form of the structures generated. The language of fractals in its present state of development seems unequal to the task of answering these questions.

Nevertheless, it seems that the study of statistical processes characterised by an ultrametric will enable some progress to be made. As Figure 6.2 shows, ultrametric structures can be associated with fractal sets: if E is a tree structure and φ_n^{n-1} the function that gives the "father" of the "children" of order n, then for all $k \leqslant n$

$$\varphi_n^k = \varphi_{k+1}^k {}_0\varphi_{k+2}^{k+1} {}_0 \cdots {}_0\varphi_n^{n-1} \tag{6.2}$$

The set $\{E_n, \varphi_n^k\}$ is a projective system such that $E = \lim \{E_n, \varphi_n^k\}$ is the Cantor set; the distance $l(s, t)$ along the tree between two points s, t of E satisfies the ultrametric inequality $l(s, t) \leqslant \max [l(s, u), l(u, t)]$.

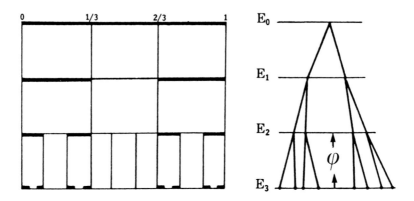

Figure 6.2 Principle of the construction of a graph that defines an ultrametric.

LARIGALDIE S., doctoral thesis, Université de Paris Sud, Orsay, 1985.
NIEMEYER L., PIETRONERO L. and WIESMAN, *Phys, Rev. Lett. 52*, 1984, 1033.

6.6 Combinatorial tree structures and stochastic matrices

The trees of Plate 4 exist only a computer screen. All were generated by the same computer program, that is, by the same algorithm, an iterative process that starts with a given set of about twenty numbers. Some deep mathematics is involved here, which is given in the references at the end of this section; the important point to note is that the form grows with time as measured by the clock.

Here we are in the field of image synthesis, which of course is something of which the computer is capable – all that is needed is the program. Algorithms for generating elementary forms such as polygons and circles are simple, but to generate forms seen in the natural world, such as clouds, mountains, landscapes or living trees, is a much more complex matter.

For practicality, the algorithm for generating a natural form, a tree for example, should have as few parameters as possible. The trees reproduced in the photographs have thousands of branchings and branches. For simplicity, each branch can be specified as a rectangle and therefore by a length and a breadth, and at each branching two angles must be given. Thus tens of thousands of

Plate 4. Combinatorial tree structures generated by stochastic matrices (6.6). (X.G. Viennot, University of Bordeaux 1) N. Janey and D. Arquès, University of Franche-Compté. Reproduced by courtesy of X.G. Viennot).

numbers are needed to describe the structure of the tree, and to these must be added more tens of thousands to describe the leaves. But in fact these trees were generated by giving the algorithm a matrix of only about 10 to 20 numbers: the difficulty lies in achieving such economy of means.

It is quite remarkable that so small a set of numbers contains all the information needed to convey the visual appearance of the tree – whether bushy, coniferous, thorny and so on: in a sense, this set plays the role of the DNA in our genes. A certain amount of controlled randomness is involved: from the same input parameters the algorithm can generate millions of different trees, but all will be of the same "species" – pines, oaks, birches or others: different species cannot be confused, and the form gives an idea of the parameters used to generate it. Thus the image represents both the input matrix and the part played by chance – in a way, the balance between chance and necessity.

This iteration gives rise to some sophisticated developments in the field of combinatorial (that is, discrete) mathematics. Some of the underlying ideas are due to hydrogeologists, for example to Horton and Strahler in their work on river basins, of which the analysis involves combinatorial parameters having very elegant mathematical properties; and deep arithmetical theorems have appeared in the course of the work. These concepts have been met also in information-theoretical studies, for example of the optimisation of the number of registers needed in the computation of a numerical expression.

Every natural tree, and more generally every ramified structure such as a lightning flash, a lung or a river basin, has an underlying mathematical tree. What is less to be expected is that an understanding of such trees is now seen to be necessary for the understanding of the secondary structure of single-strand amino acid chains such as in RNA, transfer- and messenger-RNA, etc, and for making predictions concerning properties of such structures. In France, J. Vannimenus has applied theories of this type to the analysis of ramified structures associated with fractal geometry.

Optimisation of entropy functions on these trees still remains to be undertaken; but the reader will find some first steps in this direction in econometric studies made by the present author.

GREEN M. and SUN H., *IEEE Comp. Grap. and Appl.* 8(6), 1988, 52.
FLAJOLET P., RAOULT J.C. and VUILLEMIN J., *Theoretical Comp. Sci. 9*, 1979, 99.
LE MÉHAUTÉ A., *Ann. du Centre de Recherche de L'Urbanisme*, 1974, 138.
LE MÉHAUTÉ A. and APPELBY J., *Energy 2*, 1977, 105.
PEROCHE B., ARGENCE J., GHAZANFARPOUR D. and MICHELUCCI D., *La synthèse d'images*, Hermès, Paris, 1988.

PIXIM 89, *Actes,* edited by Gagalowicz, Hermès, Paris, 1989.

PRUSINKIEWICZ P. and SANDNESS G., *IEEE Comp. Graph. and Appl. 8(6),* 1988, 26.

SMITH A.R., *Comp. Graph. 18(3),* 1984, 1.

STRAHLER A.N., *Bull. Geo. Soc. Ame. 63,* 1952, 1117.

VANNIMENUS J., *Proc. Summer School in Theor. Phys.,* Les Houches, 1987.

VANNIMENUS J. and VIENNOT X.G., *J. Stat. Phys.* 54 (5/6), 1989, 1529.

VAUCHAUSSADE de CHAUMONT M. and VIENNOT X.G., "Lectures Notes in Biomathétics", 57, edited by V. Capasso, E. Grosso and S.L. Paven-Fontant, Springer-Verlag, Berlin, 1985, 360.

VIENNOT X.G., EYROLLES G., JANEY N. and ARQUES D., Siggraph' 89 Conf. in *Comp. Graph. 23(3),* 1989, 31.

VIENNOT X.G., "Trees" in *Mots,* in press, Hermès, Paris.

6.7 Trees in speech studies: a graphical tool for language

Speech processing is a very difficult field, at the frontier between advanced signal processing and artificial intelligence. There is a lack of tools to enable the results of analysis to be visualised; we shall show here that the representation of a tree structure can provide the basis for such a tool, and that the use of this can develop skills similar to those of the botanist.

Briefly, speech is an acoustic signal consisting of a number of stable elementary zones, a structure that suggests representation by a linear finite-state automaton, each state of which is associated with one of these zones. Such an automaton can be modelled by a Markov chain such as Figure 6.3, as is commonly done in speech modelling. Here the probability of remaining in a particular acoustic zone is that associated with the closed-loop transition from that state back to itself; and of moving from one zone to the next is that of the transition between consecutive states.

Figure 6.3 State transitions based on a stochastic matrix.

Plate 5. Tree structures corresponding to spoken words (6.7). Tree D, given for comparison, was generated by a Markov chain unrelated to any word.

In this way the set of possible pronunciations of a given word can be modelled by a Markov chain, and such a model can be represented by a square matrix of transition probabilities. The diagonal elements of the matrix are the loop probabilities $a_{a,i}$, the band above the diagonal contains the transition probabilities $a_{i,i-1}$ and all the remaining elements are zero; the sum of the elements in each row is 1.

Any path along the Markov chain for a given word, from the initial state to the final, corresponds to a possible sequence of acoustic events that represents that word. The set of possible paths can be represented by a tree: an elementary tree structure can be associated with each type of transition.

The depth of the tree, called the Strahler number, is the number of states in the chain. If we start from the root and draw numbers at random we can go from a tree structure of order k (i.e. Strahler number k) to one or order l by travelling along the chain from the initial state $i = k$ to the final state $i = l$. If the loop probabilities are small this is achieved quickly, but if they are large the tree becomes bushy and spreads widely. When the value l is reached the simple model, as is the case here, scatters a number of points in its neighbourhood to represent the leaves.

In Plate 5 the photographs A, B, C show three words modelled by Markov chains of 6 states; D is an artificial chain that does not represent any word.

This type of graphical tool can be exploited further. The general form of the tree is controlled by a number of parameters, the length, breadth, colour etc of the branches, whose values depend mainly on the order i of the branch, $1 < i < k$. Correspondingly, for each state in a Markov chain there is an associated a set of parameters that characterises the acoustic events of that state. A suitable correspondence between these two sets of parameters will enable words of different classes to be represented by trees of different species.

The prospects opened up by this approach were first indicated by C. Godin and Y. Gedon in an unpublished report of the CGE Marcoussis Laboratories.

FOURNOL D., GODIN C., GUÉDON Y. and RICHARD P., "The SPIN continuous speech recognition system", *Proceedings of EUROSPEECH*, Paris, 1989, 84-87.
SABAH G., *L'intelligence artificielle et le langage, processus de compréhension, vol. 1, représentations des connaissances, vol. 2*, Hermès, Paris, 1989.

6.8 Creation of a fractal object by diffusion

On the atomic scale diffusion in a solid can be modelled by the random walk of a

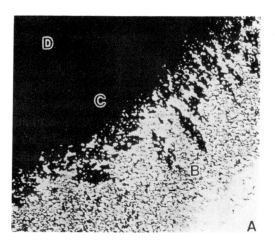

Plate 6. Solid-solid diffusion: diffusion boundary of the ceramic $Ba_2Y(Cu_{1-x}Pt_x)_3O_{6+x}$ into platinum (6.8). (J.R. Verkouteren, *Mat. Lett. 8(1–2)*, 1989, 59).

particle on a network, cf. Plate 6; and whilst on the macroscopic scale the process is described by well-known laws, e.g. Fick's Law, more complex behaviour can be expected on the atomic scale, resulting from the optimisation of the boundary, that is, production of a boundary for which the entropy is minimised.

Sapoval and his colleagues at the École polytechnique have made numerical simulations in this field which have led them to suggest that a diffusion boundary is a fractal object with, in the 2-D case, dimension $\Delta = 1.75$; this is equal to $1 + 1/v$ where $v = 4/3$ is the critical correlation length at the threshold of percolation. Their work gives also the classical relation $x_f \sim \sqrt{(D_{iff}t)}$ for the mean distance from the source to the boundary and the breadth of the boundary as $\sigma_f \sim (D_{iff}t)^{\alpha/2}$ where $\alpha = v/(1+v)$. It shows further that the structure of the boundary fluctuates violently in time.

We can arrive at the same results by a theoretical argument based on the fractional derivative, and using some of the developments of the first four chapters.

Let Δ be the dimension of the diffusion boundary; then by equation 4.63 there is a hyper-scale index β such that

$$1/d_\omega = 1/\beta + 1/\Delta - 1 \tag{6.3}$$

where $1/d_\omega = 1/2$ since the process is diffusive; thus $1/\beta + 1/\Delta = 3/2$.

This is the macroscopic view of the process, in which β expresses the fact that the interface is accessible only via a fractional Brownian motion. From the macroscopic point of view this motion parametrises, in Tricot's sense, the transfer across a fractal interface of dimension Δ, which requires the scaled decay law to be of fractance (dim $= \Delta/\beta$) type (cf. equation 2.49) with $C(p) \sim (1/p)^{\beta/\Delta}$. Locally therefore, by the laws of hyperfractality,

$$1/d_a = (\beta/\Delta) + \chi \tag{6.4}$$

The simulation shows that $1/d_a \cong 1$, so $\chi = 1 - \beta/\Delta$. Now χ is related to the dissipation in the neighbourhood of the interface and is therefore determined by the co-dimension, here $2 - \Delta/\beta$: so $1 - \beta/\Delta = 2 - \Delta/\beta$, whence $\Delta/\beta = (1 + \sqrt{5})/2$. Then since $\Delta = 1 + 1/v$ and, from (6.3), $\Delta/\beta = 3\Delta/2 - 1$ we get finally $v = 3/\sqrt{5} \cong 4/3$.

The exponent of parametrisation of the curve, $1/\beta = 0.927$, is a dimension of the fractional Brownian motion associated with access by the particles to the interface equal to $\beta = 1.078$, which leads to a dimension of the thick structure of the boundary equal to $2/\beta = 1.854$.

The agreement between the result obtained theoretically in this way and that given by the simulation suggests that theoretical solutions, based on the analysis of dissipation, could be found to some simple problems of morphogenesis.

Creation of a fractal interface by diffusion

BUNDE A. and GOUYET J.F., *J. Phys. A: 18*, 1985, L285.

SAPOVAL B., ROSSO M. and GOUYET J.F., *J. Phys. lett. 46*, 1985, L149.

SAPOVAL B., ROSSO M., GOUYET J.F. and COLONNA J.F., *Solid State Ionics 18-19*, 1986, 21.

SAPOVAL B., ROSSO M. and GOUYET J.F., "Fractal Interface and Diffusion, Invasion and Corrosion" in *The Fractal Approach to Heterogeneous Chemistry*, edited by D. Avnir, J. Wiley and Sons, New York, 1989.

Fractography

GIBOWICZ S.J., *Pure Appl. Geophys. 124*, 1986, 11.

ISHIKAWA K., OGATA T. and NAGAI K., *J. Mat. Sci. 8*, 1989, 1326.

LE MÉHAUTÉ A., *Abstract of Mat. Res. Soc.*, Fractal session of Boston Meeting, 1986, 17.

MANDELBROT B., PASSOJA D.E. and PAULLAY A.J., *Nature 308*, 1984, 721.

PASSOJA D.E., "Fractography of Glass and Ceramics", *Advances in Ceramics 22*, edited by American Ceramics Society, Inc, 1988, 101.

TERMONIA Y. and MEAKIN P., *Nature 320*, 1986, 429.

6.9 Tribology or geography?

Recent work on self-affine graphs has led to the possibility of representing naturally-occurring surfaces. As a first approximation, the altitude $z(x, y)$ at the point (x, y) has the scale relation

$$z(bx, by) = b^H z(x, y) \qquad (6.5)$$

for all values of b; H is the co-dimension, which for a surface in three dimensions is $3 - \Delta$.

With the practical limits to the frequencies the ruggedness of the surface can be generated by summing spectral components $C(v_n) \exp(-2\pi v_n x)$, so that, for example, the profile of a vertical section is given by

$$z(x) = \Sigma_n C(v_n) \exp(-2\pi v_n x) \qquad (6.6)$$

Figure 6.4 is an example of the type of profile that can occur naturally.

If such a function is to obey a scaling law of the type of equation (6.5) then the spectral power density must also obey a law of this type, i.e. $C(v_n) \sim v_n^{-H}$. Sections generated by means of these relations have a natural look and as Plate 7 shows, surfaces formed by assemblages of such sections give the appearance of contorted geographical structures.

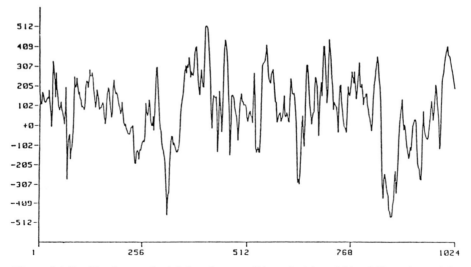

Figure 6.4 Profile of a rough nickel surface sandblasted with carbide of 60-mesh particle size.

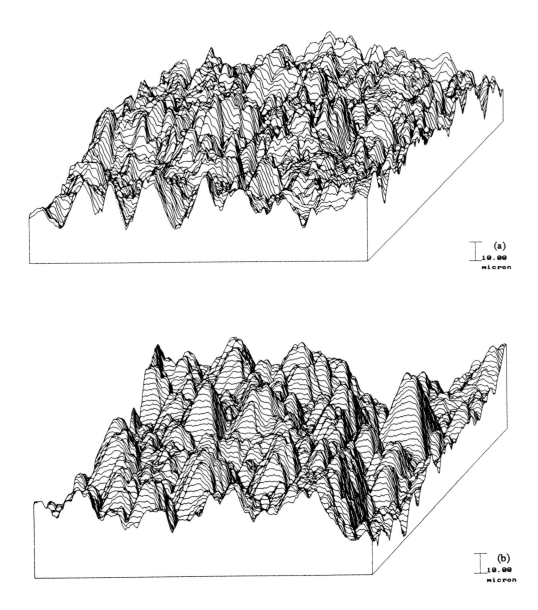

Plate 7. (a) Contour map of sandblasted brass surface derived from scanning microscope measurements. (b) Numerical simulation of (a) based on an evaluation of the fractal dimension by the variations method (6.9/2.2.2). (Photographic credit: Laboratory for Microanalysis of Surfaces, ENSMM, Besançon).

FEDER J., *Fractal*, Plenum, New York, 1988.

MANDELBROT B.B., *The Fractal Geometry of Nature*, Freemann, New York, 1983.

WEHBI D., *Approche Fractale de la Rugosité des Surfaces et Implications Analytique*, Doctoral Study, Université de Franche-Comté, Besançon, 1986.

WEHBI D. and ROQUES-CARMES C., in *Approche fractale des surfaces et conséquences analytiques*, Edition Eyrolles, 1988, 330.

GAGNEPAIN J-J. and ROQUES-CARMES C., *Wear, 109*, 1986, 119.

TRICOT C., WEHBI D., QUINIOU J.F. and ROQUES-CARMES C., *Acta-Stereologica 6/III*, 1987, 839.

DUBUC B., ROQUES-CARMES C., TRICOT C. and ZUCKER S.W., *SPIE Visuals Communications and Image Processing II, 845*, 1987, 241.

TRICOT C., ROQUES-CARMES C., QUINIOU J.F., WEHBI D. and DUBUC B., *Rev Phys. Appl. 23*, 1988, 111.

ROQUES-CARMES C., WEHBI D., QUINIOU J.F. and TRICOT C., *Surface Topographie, 1*, 1988, 237.

TRICOT C., WEHBI D. and ROQUES-CARMES C., "Perturbation et propriétés höldériennes en traitement du signal" and "Caractérisation des répartitions coplanaires d'agrégats par l'indice de Besicovitch-Taylor", in press, *Physique Appliquée*.

Chapter 7

Fractal geometry and irreversibility

Whilst the relation between irreversibility and morphogenesis is effectively unknown, irreversibility on a fractal support is now fairly well understood. It is characterised by the existence of entropy and in this chapter we outline some of the problems concerning entropy in a fractal medium.

7.1 Adsorption on a fractal object: chromatography

The background is that chromatography is based on molecular adsorption on a porous material.

A measure of the content of a dense set Γ is provided by the Minkowski-Bouligand (MB) index Δ_{MB} (§ 2.1.2) in the form $A \sim \eta^\Delta$ where $\Delta = \Delta_{MB}$ is given by equation 2.15. The physical problems to which this index is relevant must satisfy two conditions:

1. There must be a "transfer function" associated with the set Γ, for example the phenomenon must depend on the length of an interface represented by Γ when this is approximated with a step of length η.
2. They must involve dissipative phenomena, or variables associated with thickness of the boundary of Γ.

The second condition is very difficult to achieve since the physical content of an interface is usually different on the two sides – e.g. metal/vacuum, oxide/ electrolyte. If particular cases such as δ-transfer (the TEISI model) are excluded, the interface must be regarded as asymmetric and approached either from one side or the other; the question then arises of which index to use.

This is a difficult problem which, mathematically, is still unsolved and concerning which only a few short remarks can be made. Suppose A1. A2 are open sets with a common boundary dA; let ∂A be set of limit points of the sequence that approximates the points of A, that is, the limit of the sequence of different internal and external coverings by balls of diameter η, cf. Figure 7.1. To answer the question concerning the index we define the difference set $E = \partial A - da$. The external dimension $\delta 1$ of E is such that as $\eta \to 0$

$$\eta^{\delta 1 - 2} |\partial A|_2 \to 0, \quad \text{i.e.} \quad \delta 1 = \lim [2 - (\log |\partial A|_2)/(\log \eta)]$$

It can be shown that

$$\delta 1 = \Delta_{\mathrm{MB}} \quad \text{if} \quad E = 0, \neq \Delta_{\mathrm{MB}} \quad \text{if} \quad E > 0 \qquad (7.1)$$

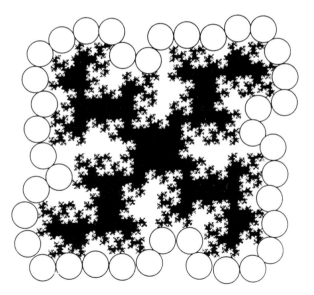

Figure 7.1 The question raised by adsorption is, which index controls molecular adsorption at a fractal interface? Clearly this must be related in some way to the dimension of the neighbourhood of the interface, that is, to the co-dimension; but certain conditions must be satisfied, as stated in §7.1. The figure illustrates the way in which the balls can be located within the boundary and the limiting interaction between balls of decreasing size (see text).

Returning to the adsorption problem, this means that a necessary condition for the relevance of the Minkowski-Bouligand index is that the residual set E is of measure zero; and if this is so then it is immaterial whether Γ or its neighbourhood is measured. In particular, if Γ is regular (see § 2.1.3) we can put $D_f = \delta 1 + 1$, so $2 - \delta 1 = 3 - D_f$; and a fractal concentration can be defined for each neighbourhood of Γ by $C_f(E) = M/|\partial A|_2 \sim M\eta^{Df-3}$, a relation that provides a mathematical explanation of many results obtained in chromotography.

NB: The condition that E is of measure zero applies equally to equation 2.51; counter-examples in which this does not hold are easily constructed.

AVNIR D., FARIN D. and PFEIFER P., *J. Chem. Phys. 79*, 1983, 3566; *Surf. Sci.126*, 1983, 59; *Nature 308*, 1984, 261.

BROCHARD-WYART F., GHAZI A., LE MAIRE M. and MARTIN M., *Chromatographia 27(5/6)*, 1989, 257.

BROCHARD-WYART F., *C.R. Acad. Sci. Paris 304 II(14)*, 1987, 785.

CHENG E., COLE M.W. and PFEIFER P., *Phys. Rev. B39(17)*, 1989, 12962.

COLE M.W., HOLTER N. and PFEIFER P., *Phys. Rev. B 33(12)*, 1986, 8806.

PFEIFER P. and AVNIR D., *J. Chem. Phys. 79*, 1983, 3558.

PFEIFER P. and SCHMIDT D.M., *Phys. Rev. Lett 60(13)*, 1988, 1345.

PFEIFER P., *Chemistry and Physics of Solid Surface VII*, edited by R. Vanselow and R.F. HOWE, Springer-Verlag, Berlin, 1988, 283.

PFEIFER P., KENNTNER J. and COLE M.W., "Fundamentals of Adsorption" in press, *Proc. of Conf.*, Sonthofen, 1989.

PFEIFER P., OBERT M. and COLE M.W., *Proc. R. Soc. Lond. A 423*, 1989, 169.

PFEIFER P., YU Y.J., COLE M.W. and KRIM J., *Phys. Rev. Lett 62(17)*, 1989, 1997.

TRICOT C., *Phys. Lett.14 A(8/9)*, 1986, 430.

TRICOT C., "Dimensions aux bords d'un ouvert", personal communication.

7.2 Thermodynamics of curves, entropy and dimension

Fractality is not a necessary condition for complexity: a knot is a complex curve but is derivable almost everywhere. Mandès-France, who among others has done research on complex rectifiable plane curves, has shown that for a large class of curves Γ, an *entropy* h can be defined such that $h \leqslant 1 - 1/\dim \Gamma$; we define h below and show how this inequality can be deduced from an approximation to the curve of type MB

Let $\Delta_{inf} = \inf[(\log|\Gamma|^2)/\log R]$, $\Delta_{sup} = \sup[(\log|\Gamma|^2)/\log R]$, where $|\Gamma|^2$ is the area enclosed within a ball of radius R of a corridor of width η centred on Γ, for $R \to \infty$; let $\dim_{inf}(\Gamma) = \lim \Delta_{inf}$, $\dim_{sup}(\Gamma) = \lim \Delta_{sup}$ as $\eta \to 0$. Γ is said to be *resolvable* if $\dim_{inf}(\Gamma) = \Delta_{inf}$ and $\dim_{sup}(\Gamma) = \Delta_{sup}$, which means that it is immaterial whether the resolution of the curve or the size of the domain is increased. It is *unresolvable* if $\dim_{inf}(\Gamma) < \Delta_{inf}$ and $\dim_{sup}(\Gamma) = \Delta_{sup}$, meaning that the curve would still be seen as chaotic even if viewed through a microscope of infinite resolving power.

Consider now the intersection of Γ with an arbitrary straight line; the number N of intersections is

$$N = \sum_n n\, p_n = 2L/F \tag{7.2}$$

where p_n is the probability of n intersections, L is the length of Γ and F is the length of a connected envelope (see Fig. 7.2).

For any finite curve $F \leqslant 2L$, and the entropy in the Shannon sense can be expressed as

$$S(\Gamma) = \sum_n p_n \log(1/p_n) = -\sum_n p_n \log p_n \tag{7.3}$$

so that a highly contorted curve will have a high entropy. A classical variational calculation gives $p_n = \exp(-\beta n)$ where, assuming that L is finite, $\beta = \log[2L/(2L - F)]$, the "thermodynamic temperature" of the curve; so for any finite curve $S(\Gamma) \leqslant \log(2L/F) + \beta/(e^\beta - 1)$.

Mendès-France extends this to the case of a infinite curve by defining

$$h_{sup}(\Gamma) = \sup[S(\Gamma)/\log|\Gamma|^2], \quad h_{inf} = \inf[S(\Gamma)/\log|\Gamma|^2] \tag{7.4}$$

In favourable cases

$$h_{sup} = h_{inf} = h(\Gamma) \tag{7.5}$$

and in all cases

$$0 \leqslant h(\Gamma) \leqslant 1$$

The curve is deterministic if $h = 0$, chaotic if $h = 1$; and the entropy is related to the dimension by

$$h(\Gamma) = 1 - 1/\dim(\Gamma) \tag{7.6}$$

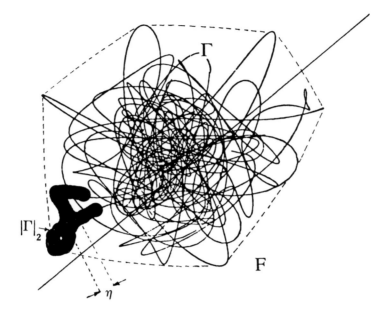

Figure 7.2 Entropy of a curve. Simulation of the transmission of an acoustic signal, frequency 406 Hz, over 1000 km of ocean: phase-amplitude diagram in the complex plane. J. Prendry, *Nature* 342, Nov 1989, 223.

Suppose $h(\Gamma) = 1/2$; then $1/2 \leqslant h(\Gamma) \leqslant 1 - 1/\dim(\Gamma)$; so $\dim(\Gamma) \geqslant 2$, which is impossible for an ordinary plane curve. The curve therefore must have an infinite number of double points.

NB: The Mendès-France papers should be read; they explain the relations between fractal curves and give a transformation that is important for the analysis of irreversibility.

DAVIS C. and KNUTH D., *J. Recreational Math.* 3, 1970, 61 and 133.

DEKKING F.M. and MENDES-FRANCE M., *J. Für die Reine und Angewandte Mathematik 329*, 1981, 143.

DEKKING F.M. and MENDES-FRANCE M., *Mathematical Intelligencer 4*, 1982, 130, 173 and 190.

MENDES-FRANCE M. and TENENBAUM G., *Bull. Soc. Math. Fran. 109*, 1981, 207.

MENDES-FRANCE M., *Bull. Austral. Math. Soc. 24*, 1981, 123.

MENDES-FRANCE M., DUPAIN Y. and KAMAE T., *Arch. for Rational Mechanics and Analysis 94*, 1986, 155.

MENDES-FRANCE M., *Physics Reports 103*, 1984, 161.
MENDES-FRANCE M., "Images de la physique", *Le courrier du CNRS*, 51, 1983, 5.
STEINHAUS H., *Colloquium Mathematicum 3*, 1954, 1.

7.3 Chemical potential, entropy and dimension

A statistical treatment of fractality shows very clearly the relation between entropy and fractal dimension. Entropy is a thermodynamic property, measurable in all chemical processes in which the free enthalpy of the reaction is constant; for then, provided that the temperature is constant, the free energy is proportional to the entropy. Also, the fractal dimension of the thermodynamic object being studied can be found by harmonic analysis of the transfer or state function (cf. §§ 2.3.4, 7.4) and thus a relation between dimension and thermo-

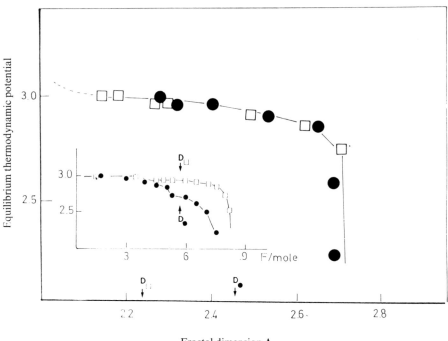

Figure 7.3 The inset gives the chemical potential of lithium in two $Li_x MnO_2$ materials, as a function of x. A fractal dimension can be associated with each point (cf. § 7.4) and a plot of the potentials against these gives the single curve shown.

dynamic potential can be determined experimentally. This has been done for the penetration of lithium into allotropic matrices of manganese dioxide to give Li_xMnO_2, for which both the chemical potential and the fractal dimension are related to the amount of penetration; Figure 7.3 gives some results, entropy here being represented by thermodynamic potential.

The experimental circumstances giving such a relation are exceptional, since usually the free enthalpy is not constant and therefore the potential that is measured is not determined by a single geometry. This leads to the need to consider multifractality, for which the papers by Mandelbrot should be consulted.

LE MÉHAUTÉ A. and DUCAST A., *Journal of Power Sources*, 9, 1983, 359.
LE MÉHAUTÉ A., "Fractal Electrode", in, *The Fractal Approach to Heterogeneous Chemistry*, edited by D. Avnir, J. Wiley and Sons, New York, 1989.
SANDRE E. and LE MÉHAUTÉ A., *Entropie d'état sur un ensemble fractal*, in press.

7.4 Time in fractal geometry

Quadratic forms, such as energy, play a central role in physics, expressing the invariance of the laws through the duality between the object of the experiment (the effects) and the environmental variables (the causes). This is why so much of elementary physics is based on proportionality between causes and effects – more precisely, between fluxes $J(t)$ (extensities) depending on time, the effects, and the forces $\Delta X(t)$ (intensities) which produce them; here $X(t)$ is a thermodynamic potential and $\Delta X(t) = X_\infty(t) - X(t)$ where $X_\infty(t)$ is a real or virtual stationary state used as a reference. Thus $J(t) \sim \Delta X(t)$.

An extensity is a variable evaluated in space, and a flux, which is expressed naturally as "per unit area", is always measured across an interface, most often implicitly assumed Euclidean. In the simple but fairly general case in which the instantaneous state function at the interface is capacitative – that is, when the quantity $Q(t) \sim \int J(t)\,dt$ of the extensity accummulated there is proportional to $X(t)$ – the dynamics of the process is expressed by a first-order differential equation:

$$dX(t)/dt \sim X_\infty(t) - X(t) \qquad (7.7)$$

This gives the well known exponential behaviour, $X(t) \sim \exp(-t/\tau)$ where τ is a time constant, shown by every deterministic process in a complete, continuous, linear normed space of dimension 1. The Laplace transform $Z(p)$ of the solution is the transfer function of the dynamics of the process, $Z(p) = X(p)/X_\infty(p) \sim 1/(1+\tau p)$, where p is the Laplace generalised frequency.

It can be shown that in a fractal medium the same assumptions lead to differential equations of non-integral order, α:

$$\mathscr{D}_t^\alpha X(t) \sim f[X_\infty(t), X(t)] \tag{7.8}$$

and to two important types of transfer function, called respectively Cole-Cole (CC) and Davidson-Cole (DC):

$$Z_{CC} \sim 1/[1+(\tau p)^\alpha], \quad Z_{DC} \sim 1/(1+\tau p)^\alpha \tag{7.9}$$

The geometrical interpretation of these relations is given by the TEISI model, which is the physical expression of the parametrisation of the fractal geometries (cf. §2.3.4). The two modes of behaviour are not independent in practice and can be related, at least at high frequencies, by basing the treatment on circuit theory. Thus regarding $L_\alpha(p)$ in Figure 7.4 as a "black box" it can be shown that, for frequencies p greater than some critical value p_{crit}, the behaviour is of type DC if the switch is open and CC if it is closed.

The parameter α is usually a simple function of the fractal dimension or co-dimension, $\alpha = 1/\Delta = 1/(D-1)$ in 2D, or the co-dimension, $\alpha = 2-\Delta = 3-D$ (cf. §2.1.3, eqn. 2.21).

Many examples of different behaviour are given in the literature of electro-chemical, dielectric, magnetic and mechanical systems, and of what is called "stretched exponential" time-dependence:

$$X(t) \sim \exp[-(t/\tau)^\beta] \tag{7.10}$$

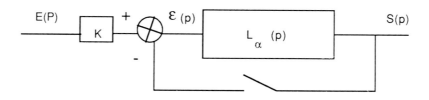

Figure 7.4 Circuit to illustrate the relation between Cole-Cole and Davidson-Coles types of behaviour.

The Laplace transform of this function is numerically very close to Z_{DC}, giving a relation between α and β: this is currently being investigated.

The importance of this analysis is that it makes it possible to distinguish between the kinetic parameters such as τ (related to the physico-chemical character of the interface) and the geometrical parameters α, β (usually related to the metric of the space in which the process occurs); and this gives possibilities for optimisation, since the two types of parameter are then accessible independently.

The advantage is particularly significant in electromagnetism, where the analysis provides a simple interpretation of electromagnetic impedance in a homogeneous medium with constant properties μ, σ, ε: if a voltage E(t) produces a current H(t) the impedance is Z(t) = E(t)/H(t) and

$$Z(p) \sim [(p\mu/\sigma)/(1+p\varepsilon/\sigma)]^{1/2} \tag{7.11}$$

This equation is easily interpreted in the framework of the model. It gives the balance between the diffusion of a current H(t) – which is a transfer across an interface of dimension 2 (a Peano curve) – and a decay of charge controlled by the transfer function of type Z_{DC}, in the neighbourhood of this curve. It follows that

$$Z_{DC}(t) * \mathscr{D}_t^{1/2} H(t) \sim E(t) \tag{7.12}$$

where $Z_{DC}(t)$ is the inverse transform of $1/\sqrt{(1+p\varepsilon/\sigma)}$, giving the Davidson-Cole type of behaviour.

The relation (7.11) can be generalised to

$$Z(p) \sim (pk_1)^{1/d1}/(1+pk_2)^{1/d2} \tag{7.13}$$

This provides a general means for studying electromagnetic impedance, by distinguishing the geometrical parameters d1, d2 from those expressing the physical-chemical properties, μ, σ, ε: see Plate 8. The experimental feasibility has been shown by work now in progress on the interaction of electromagnetic radiation with fractal media.

Oustaloup and his colleagues at the University of Bordeaux-Talence have extended the fractional-order differential equations to equations of complex order, and the results will be published shortly. It is easily shown that the complex variable results in a time-dependence in the constants of the equation, which then become susceptible to scaling laws. Oustaloup's group are now investigating the physical significance of operators of this type.

NB: The work of the present author and his group has all been concerned

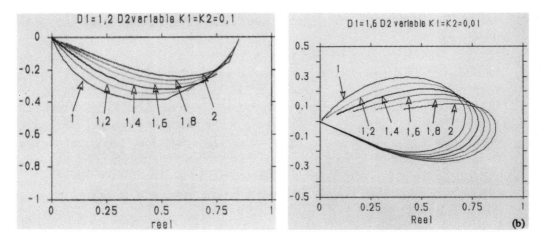

Plate 8. Electromagnetic impedance of a heterogeneous material. d_1 and d_2 are the geometrical parameters, k_1 and k_2 are the physico-chemical constants (7.4, eq. 7.13). (a) purely capacitative behaviour (b) inductive-capacitative behaviour.

with values $\alpha = 1/\Delta$ and $\alpha = 2 - \Delta$. However, other values are easily found by departing from the basic model: see for example the formulations based on a generalised co-dimension given in M. Keddam & H. Takenouti, "First International Electrochemical Symposium Spectroscopie", 22-26 May 1989, Bombannes-Maubuisson, Extended Abstract C2-8.

COLE K.S. and COLE R.H., *J. Phys. Chem.*, *10*, 1941, 98.

GEMANT A., *physics 7*, 1936, 311.

HAMAIDE T., GUYOT A., LE MÉHAUTÉ A., CRÉPY G. and MARCELLIN R., *J. Electhem. Soc.*, *136, 11*, 1989, 3152.

KEDDAM M. and TAKENOUTI H., *C.R. Acad. Sc. Paris 302(II)*, 1986, 281.

KEDDAM M. and TAKENOUTI H., *Electrochim. Acta 33*, 1988, 445.

LE MÉHAUTÉ A., DE GUIBERT A., DELAYE M. and FILLIPI C., *C.R. Acad. Sci. Paris, t. 294(II)*, 1982, 865.

LE MÉHAUTÉ A. and CRÉPY G., *Solid State Ionics, 9&10*, 1983, 17.

LE MÉHAUTÉ A., *J. Stat. phys. 36 (5&6)*, 1984, 665.

LE MÉHAUTÉ A., PICARD L. and FRUCHTER L. *Phil. Mag.B,52(6)*, 1985, 1071.

LE MÉHAUTÉ A., CRÉPY G. and HURD A., *C.R. Acad. Sci. Paris, 306*, 1988, 117.

LE MÉHAUTÉ A., "Fractal Electrode" in *The Fractal Approach to Heterogeneous Chemistry*, edited by D. Avnir, J. Wiley & Sons, New York, 1989, 311.

LE MÉHAUTÉ A., HÉLIODORE F. and CHABASSIER G., "Electromagnetic Waves in

Fractal Media", *Progress in Electromagnetism Research Symposium*, Boston, July 1989.

LIU S.H., *Phys. Rev. Lett. 55 529*, 1985.

MULDER W.H. and SLUYTERS J.H., *Electrochim. Acta 33*, 1988, 303.

NYIKOS L. and PAJKOSSY T., *Electrochim. Acta 30*, 1985, 1533.

OLDHAM K.B. and SPANIER J., *The Fractional Calculus*, Academic Press, New York, 1974.

OUSTALOUP A., *Systèmes Asservis Linéaires d'Ordre Fractionnaire, Théorie et Pratique*, Masson, Paris, 1983.

OUSTALOUP A., HÉLIODORE F. and LE MÉHAUTÉ A., *in ASI NATO*, Series "Relaxations and Related Topics in Complex Systems", in press.

PAJKOSSY T. and NYIKOS L., *Electrochim. Acta 34*, 1989, 171.

PAJKOSSY T. and NYIKOS L., *Electrochem. Soc 133(10)*, 1986, 2061.

SAPOVAL B., *Solid State Ionics 23(4)*, 1987, 253.

SAPOVAL B., CHAZALVIEL J.N. and PEYRIERE J., *Phys. Rev.A, 38*, 1988, 5867.

SCHEIDER W., *J. Phys. Chem.79(2)*, 1975, 127.

7.5 Energy distribution and fractional derivation

The fractional derivative is defined by the integral

$$_a\mathscr{D}_t^\alpha f(t) = (1/\Gamma(-\alpha)) \int_a^t f(y)(t-y)^{-(\alpha+1)}\,dy \qquad (7.14)$$

Taking $f(t) = \cos\omega t$ we have

$$_a\mathscr{D}_t^\alpha \cos\omega t = \omega^\alpha[\cos(\omega t + \pi\alpha/2) - D(a)] \qquad (7.15)$$

where $D(a) = 0$ for $a = -\infty$, and for general values $= [\cos\omega t \cdot C(\omega t, -\alpha) + \sin\omega t \cdot S(\omega t, -\alpha]/\Gamma(-\alpha)$, C, S being the generalised Fresnel integrals.

If we apply an alternating current $I(t) = I_0\cos\omega t$ to a linear circuit the voltage response is $V(t) = V_0\cos(\omega t + \psi)$, where ψ is the phase angle. From (7.14)

$$_{-\infty}\mathscr{D}_t^{2\psi/\pi}I(t) = I_0\omega^{2\psi/\pi}\cos(\omega t + \psi) \qquad (7.16)$$

$$= (1/z)V(t) \qquad (7.17)$$

where $z = (V_0/I_0)\omega^{-2\psi/\pi}$

Thus the conjugated variables current and voltage are related by the operation of fractional derivation. The power dissipation is $W = V \cdot I = V_0 \cdot I_0 \cos \omega t \cos (\omega t + \psi)$ (cf. Fig. 7.5); the phase term prevents the energy balance from being established over a single period, and if the input and output energies

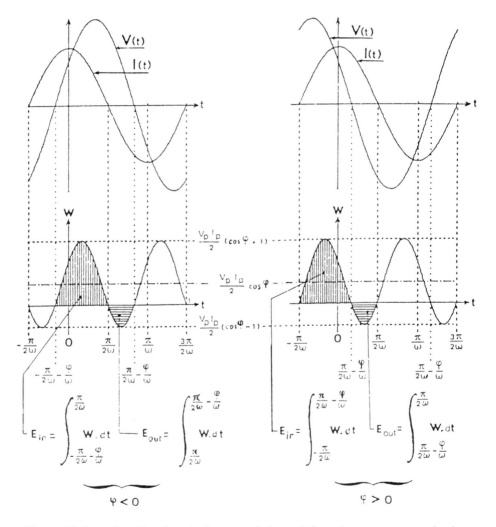

Figure 7.5 Part played by phase in the energy balance, influencing wave propagation in a complex circuit characterised by laws of scale. If the phase is related to the metric the geometry has a dominant influence on the energy dissipation (cf. Jacquelin 1986).

are E_{in}, E_{out}, and $B = V_0 \cdot I_0/2\omega$, then

for $\psi \leqslant 0$:

$$E_{in} = B[(\pi + \psi)\cos\psi - \sin\psi]$$

$$E_{out} = B[\psi\cos\psi - \sin\psi] \tag{7.18}$$

$$E_{in}/E_{out} = 1 - \pi/[\pi - \psi - \tan\psi]$$

for $\psi \geqslant 0$: $\tag{7.18}$

$$E_{in} = B[(\pi - \psi)\cos\psi + \sin\psi]$$

$$E_{out} = B[-\psi\cos\psi + \sin\psi]$$

$$E_{in}/E_{out} = 1 - \pi/[\pi - \psi + \tan\psi]$$

Thus the energy ratio depends on the phase shift and is independent of the frequency ω if and only if the phase shift also is independent of this frequency.

It seems that many linear phenomena are governed by differential equations of non-integral order; and it is seems that for these the rate of increase of entropy is distributed uniformly over the whole range of frequencies. It can be shown that this distribution minimises the entropy production by the irreversible process under consideration. The electrical circuit that leads to this property has been called *fractance* (1983), later *phasance*.

LE MÉHAUTÉ A. and CRÉPY G., *Solid State Ionics 9-10*, 1983, 17.
JACQUELIN J., *R.G.E. 1*, 1987, 47.
ROQUES-CARMES C., *R.G.E. 3*, 1987, 49.

7.6 Continued fractions, scale laws and electrical circuits

As Dieudonné remarked in his summary of the history of mathematics, the concept of continued fractions goes back to ancient China and classical Greece. To remind the reader, the idea starts by representing a positive real number x as the sum of its integer and fractional parts, $x = a_0 + \theta_0$, where a_0 is an integer (zero if $x < 1$) and $0 < \theta_0 < 1$. We can then write $1/\theta_0$ (> 1) in the same form, $1/\theta_0 = a_1 + \theta_1$, and so on, thus expressing x in the form

$$x = a_0 + 1/(a_1 + 1/(a_2 + \ldots + 1/(a_i + \ldots \tag{7.19}$$

where all the a_i, of which there may be a finite or infinite number, are positive integers.

There are many theorems concerning continued fractions. Since for an electrical circuit consisting of components in parallel the total admittance – the reciprocal of the impedance – is the sum of the admittances of the individual components these theorems can be used to study circuits, in particular to simulate types of behaviour that originate in fractality.

To take an example due to Jacquelin, consider a material for which the restivity $\rho(x)$ and capacity per unit area $\gamma(x)$ depend on a certain distance x, for example the depth of penetration into a porous material (cf. Fig. 7.6). If the fractality is expressed by power-law dependencies $\gamma(x) = \gamma_1 x^a$, $\rho(x) = \rho_1 x^b$ (see Sapoval et al 1988) the differential equation for the current I is

$$\mathcal{D}_x^{1}[x^{-a}\mathcal{D}_x^{1}(I)] = i\omega\gamma_1\rho_1 x^b I \tag{7.20}$$

With given boundary conditions the solution is

$$I(x) = I_0[2/\pi]\sin(\pi v)\Gamma(-v)(i^{3/2}\theta)^{1+v}[\mathrm{Ker}_{1+v}(\theta)+i\mathrm{Kei}_{1+v}(\theta)] \tag{7.21}$$

where

$$v = -[(b+1)/(a+b+2)] \quad \text{and} \quad \theta = [2\sqrt{(\gamma_1\rho_1\omega)}/(a+b+2)]x^{(a+b+2)/2} \tag{7.22}$$

and Ker, Kei are the Kelvin functions.

The impedance for $x \to 0$ is given by

$$Z^*_{x\to0} = V(0)/I(0) = \{(1/i\omega\gamma_1 x^a)[(1/I)\mathcal{D}_x^{1}I] \tag{7.23}$$

that is, $Z^* = P \cdot (i\omega)^v$,

$$P = [\Gamma(-v)/\Gamma(2+v)][(a+1)/(a+b+2)^{2(1+v)}](\gamma_1\rho_1)^v\rho_1 \tag{7.24}$$

Thus in the limit as $x \to 0$ the impedance of the material behaves like a fractance (or phasance), with phase angle $\varphi = -\frac{1}{2}\pi(b+1)/(a+b+2)$. In other words, in the limit of approximation the scale laws for the basic physico-chemical parameters impose scale laws on the impedance; and so with respect to measure the system just described behaves like a fractal, cf. equations 2.43, 2.45.

An analogy can be given in terms of electrical circuits. For the circuits shown in Figure 7.6 suppose the capacitances and resistances in the successive cells are given by

$$C_n = C_1 h^{n-1}, \quad R_n = R_1 k^{n-1} \tag{7.25}$$

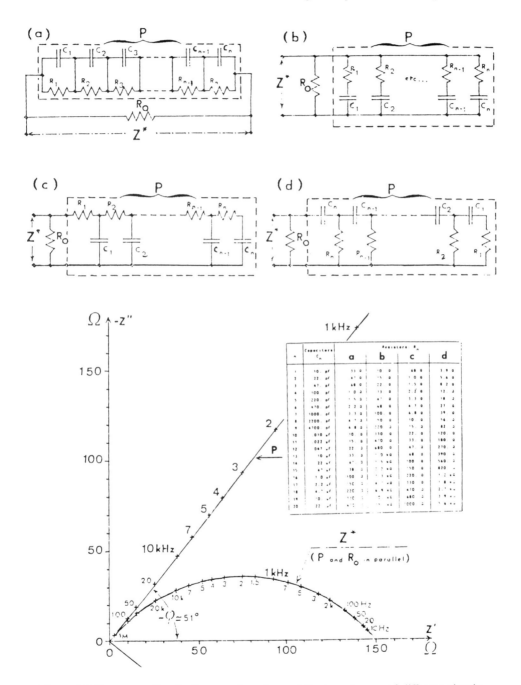

Figure 7.6 Representation in the complex plane of the impedances of different circuits having the same value of Δ (Jacquelin).

The impedance of the circuit is then:

$$Z^* = P(i\omega)^\nu$$

$$\nu = 2\varphi/\pi = 1/[1 + \log h/\log k]$$

$$P = [\Gamma(-\nu)/\Gamma(2+\nu)] [-\nu(1+\nu)R_1/k - 1]^{1+\nu}[C_1/h - 1]^\nu \qquad (7.26)$$

As the figure shows, the same scaling laws can apply to several different circuits; this is one expression of the fact that there is no 1-to-1 relation between metric (here, the scaling laws) and topology (here, the form of the circuit).

The form of the expression for ν suggests that this quantity is a generalised co-dimension. If we put $\Delta = 2 - \nu$, $N_c = h^2$, $N_r = k$ we find

$$\Delta = (\log N_c + \log N_r)/(\log h + \log k) \qquad (7.27)$$

which is a generalisation of the original Mandelbrot formula.

The bibliography gives references to many studies undertaken along these lines; that of Barnsley in particular derives Julia sets as sets of poles of iterated electrical circuits. All these studies lead to the conclusion that phasance-type behaviour is not limited to fractal systems but can be shown by any multi-branched network subject to scaling laws. This suggests an extension of the concept of dimension, for which equation (7.27) provides an example.

NB: It will be seen that any triangular circuit subject to scaling laws can be transformed into a tree of components that defines an ultra-metric, which in turn is related to a projective fractal system. This is the mathematical reason for the relation between the behaviour of a fractal circuit and the co-dimension of the fractal system that it represents.

ABRAMOWITZ M. and STEGUN S., *Handbook of Mathematical Functions*, Dover Publ., 1970, 9th edition.

JACQUELIN J., "First International Symposium on Electromechanical Spectrocopy", Bombannes-Maubuisson, France, 1989, *Extended Abstract c.2.8*.

BARNSLEY M.F., MORLEY T.D. and VIRSCAY E.R., *J. Stat. Phys.* 40(1/2), 1985, 39.

KAPLAN T. and GRAY L.J., *Phys. Rev. B32*, 1985, 7360.

KAPLAN T., GRAY L.J. and LIU S.H., *Phys. Rev. B35*, 1987, 5379.

KEDDAM M. and TAKENOUTI H., "First International Symposium on Electro-mechanical Spectroscopy", Bombannes-Maubuisson, France, 1989, *Extended Abstract c.2.8*.

LAKHTAKIA A., MESSER R., VARADAN V.V. and VARADAN V.K., *Naturforsch. 43a*, 1988, 943.

LIU S.H., *Phys. Rev. Lett. 55*, 1985, 529.

SAPOVAL B., CHAZALVIEL J.N. and PEYRIERE J., *Phys. Rev. A 38(11)*, 1988, 5867.

7.7 Fractional differential equations, Laplace equations

We consider fractional filters: these are a generalisation of the ordinary ARMA filters for which the Laplace transform of the transfer function f(t) has the form:

$$F(p) = \prod_{j=1}^{s} (1 - \alpha_j p)^{d_j} \tag{7.28}$$

The exponents d_j can be complex, say $d_j = R(d_j) + iI(d_j)$; if $\Sigma R(d_j) < 0$ then f(t) is locally integrable and gives the impulsive response of the filter; it can be shown that near $t = 0$ f(t) has the form t^k where $k = 1 + \Sigma R(d_j)$.

Several types of fractional differential equation can be associated with fractional filters. Given certain conditions, to be stated later, the following mathematical objects can be associated with any such filter:

– the transfer function F(p) and the impulsive response f(t)
– a differential equation with respect to time t
– a differential equation with respect to the Laplace variable p
– two polynomials B(p), C(p)

These are not all independent.

The polynomials B(p), C(p) are obtained from F(p) by taking logarithms and differentiating:

$$\mathscr{D}_p^{-1} \log F(p) = -\Sigma \alpha_j d_j (1 - \alpha_j p)^{-1} = B(p)/C(p) \tag{7.29}$$

Given certain simple conditions for validity, f(t) is the solution of a linear homogeneous differential equation whose coefficients are linear functions of t.

The physical significance of this mathematical result is that if a time-dependent physical phenomenon can be represented by a linear differential equation of integral order with coefficients that are affine functions of time, then it can be represented equivalently by a differential equation of non-integral order with constant coefficients. The physicist always looks for constants in his equations; the fractional order of the transformed equation and the constant coefficients are true state constants. The TEISI model is simply another example of this mathematical property.

Proceeding in a complementary manner we can look for solutions of equations of the form

$$\sum_0^N (\beta_k + \gamma_k t) y^{(k)}(t) = 0, \quad y^{(k)}(0) = 0, \quad k = 1, 2, \ldots N \tag{7.30}$$

Such equations are well known; when possible they are solved in terms of t, otherwise, after a Laplace transform, in the complex plane. The transform can be expected to give the following simplifications:

- introduction of the polynomials B(p), C(p), easily calculable given the β_k, γ_k
- representation as a fractional filter F as in (7.28).

The transform Y of y (when it exists) satisfies a homogeneous linear differential equation of order 1 whose coefficients are polynomials in p of order N or less.

A rather more complex form of equation is

$$\sum_0^N m_k(t) y^k(t) = 0 \tag{7.31}$$

where the m_k are polynomials in t of degree $r \geqslant 1$. The Laplace transform interchanges orders and degrees: the order is r and the coefficients are polynomials are of degree N. This is of interest when $r < N$; if $r \geqslant N$ it can be useful if the resulting equation is simpler than the original, or is a known form.

Finally, the stability of the solutions – which has not been studied much – seems to be related to the polynomials B and C, which therefore are important in connection with the use of fractional filters in automatic control. Their role is analogous to that of the polynomial differential operator in a linear equation with constant coefficients. This would seem to be study worth pursuing, particularly for applications in control engineering and electromagnetism; work along these lines is being done by C. Viano, A. Brami and G. Oppenheim in the Statistics Group of the Mathematical Laboratory at the University of Paris, Sud, Orsay.

7.8 Fractional derivation and diffusion

Many transport phenomena are governed by flux-force equations of the type

$$J(x, t) \sim \partial_x [\Delta X(x, t)] \tag{7.32}$$

where $\partial_x^a = \partial^a/\partial x^a$ and $\Delta X(x, t)$ is a scalar quantity. If the system is conservative and the divergence of the flux is zero and we have the *diffusion equation*

$$D_{iff}\partial_x\partial_x\Delta X(x, t) \sim \partial_t\Delta X(x, t) \tag{7.33}$$

which corresponds to transport by Brownian motion. D_{iff} is a diffusion coefficient if the process considered is controlled by a gradient of chemical potential, or in the case of heat conduction has the value $\kappa/\rho\sigma$ where κ, ρ, σ are the thermal conductivity, density and specific heat respectively.

For convenience, put $D_{iff} = k^2$; the equation is

$$(\partial_t - k^2\partial_x\partial_x)\Delta X(x, t) = 0 \tag{7.34}$$

Factorising the operator,

$$(\partial_t^{1/2} - k\partial_x)(\partial_t^{1/2} + k\partial_x)\Delta X(x, t) = 0 \tag{7.35}$$

It can be shown that only the first operator need be considered:

$$(\partial_t^{1/2} + k\partial_x)\Delta X(x, t) = 0 \tag{7.36}$$

from which, with (7.32), we get

$$\partial_t^{1/2}\Delta X(x, t) = -k\partial_x\Delta X(x, t) = -kJ(x, t) \tag{7.37}$$

This is a particular form of equation 2.48, with $\Delta = 2$, and also an expression for a Brownian process. δ-transfer models relate the coefficient $1/2$ to the homogeneous character of the space and reduce diffusion to a simple process of order 1 on a fractal interface of dimension 2.

BIRKE R.L., *J. Anal. Chem.*, *45(13)*, 1973, 2292.
LE MÉHAUTÉ A., DE GUIBERT A., DELAYE M. and FILLIPI C., *C.R. Acad. Sci. Paris*, t. *294(II)* 865, 1982.
LE MÉHAUTÉ and CRÉPY G., *Solid State Ionics, 9&10,17*, 1983.
NIGMATULLIN R.R., *Phys. Stat. Sol. (b) 133*, 1986, 425.
NIGMATULLIN R.R., *Phys. Stat. Sol. (b) 133*, 1986, 713.
NIGMATULLIN R.R., *Phys. Stat. Sol. (b) 153*, 1989, 49.
OLDHAM K.B. and SPANIER J., *J. Electroanal. Chem. 26*, 1970, 331.
OLDHAM K.B. and SPANIER J., *J. Math. Anal. 39*, 1972, 655.
OLDHAM K.B. and SPANIER J., *The Fractional Calculus*, Academic Press, New York, 1974.

7.9 Motive power and energy yield in a fractal medium

I can think of no better example than ordinary batteries to illustrate the part played by fractal geometry in influencing the effectiveness of our actions.

For any system for storing energy the space-time coupling that is inherent in fractal geometry leads to a relation between the quantity of energy stored, measured in watt-hours (Wh) and the power (measured in watts, W) that it makes available, a quantity that depends on the area of the interface with the external medium and the time over which the energy is transferred.

Suppose, for example, that a battery is discharged through a variable resistance so that the current remains constant and the voltage continues to fall until the battery is flat. The voltage/time curves for different discharge rates will resemble one another and the integral of each, multiplied by the constant current, gives the energy available. The ratio of this to the amount theoretically available is the *yield* of the battery for that discharge current.

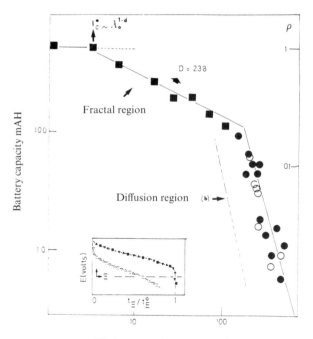

Figure 7.7 Capacity of a battery as a function of discharge rate, for various discharge-limiting regimes.

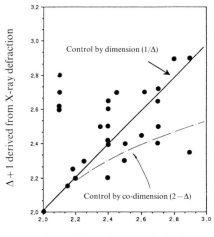

Fractal dimension $\Delta + 1$ derived from
electrochemical measurements

Figure 7.8 Comparison of two methods of measuring the fractal dimensions of samples
of carbon black: electrochemical (cf. §§ 2.43, 2.45, 7.38) and X-ray diffraction
(§ 7.2).

In the ideal case the theoretical energy is a constant, given by $VI\tau = VQ_{th}$
where $Q_{th} = I\tau$, the theoretical capacity of the battery and τ is the time for
complete discharge at a constant current I, which will depend on the *cut-off
voltage*. The capacity found in practice will usually be less than the theoretical
value and will follow the law

$$I^{\alpha}Q = C(E) \qquad\qquad (7.38)$$

where E is the *cut-off voltage* chosen by the experimenter. It can be shown that α
depends on the fractal dimension of the electrode that limits the discharge; in
general, $\alpha = D - 2 = \Delta - 1$ if the battery is in good condition (δ-transfer regime),
if the battery is subjected to a particularly heavy current. (see Figs. 7.7, 7.8).

These results are undoubtedly of practical importance. With certain
reservations, equation (7.38) is independent of E, so that a simple change of time
scale enables all the discharge curves to be superimposed on one another; and
ideally this linear transformation depends only on the fractal dimension of the
electrode.

HURD A.J., SCHREFER D.W., SMITH D.M., ROSS S.B., LE MÉHAUTÉ A. and SPOONER S. *Phys. Rev. B, 39(13)*, 1989, 9742.
LE MÉHAUTÉ A., PICARD L. and FRUSCHTER L., *Philosophical Magazine, 52*, 1985, 1071.
LE MÉHAUTÉ A. and CRÉPY G., *J. of Power Source 26*, 1989, 179.
MARCHAND A., J., *Phys. 39*, 1978, 117.
PEUKERT W., *Electrotech. Z. 18*, 1897, 287.

7.10 An example of hyper-fractality

Among the more striking successes of the TEISI model two, to my mind, are particularly relevant to our discussion. One is the generalisation of the concept of velocity, with the unification within a single framework of the concepts of diffusion coefficient, velocity and acceleration. The other is the predicting, in advance of any experimental observations, of the correlations between the high- and low-frequency dissipative processes observable in fractal media (hyperscaling).

The idea is as follows. The TEISI model is based on the existence of a relaxation process called δ-transfer, characterised by dissipation in a "Minkowski sausage" which is considered to be homogeneous. This process can

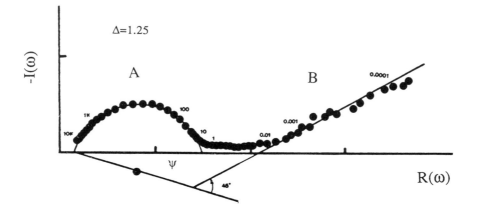

Figure 7.9 Relaxation spectrum for the diffusion of lithium ions, with the X and Y axes related to current and efficiency respectively. The efficiency is 100% up to about 2 mA and between 2 mA–200 mA decreases according to a power law with slope 1.38. At higher currents diffusion occurs in the electrode and the power law is related to a hyperscaling process.

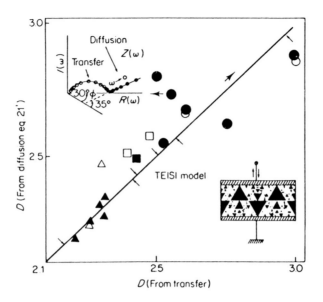

Figure 7.10 Comparison of the fractal dimensions derived from a Δ-transfer process (equation 2.39) and a diffusion process (equation 4.40) for a solid plastic electrolyte that has been aged for 5 years.

be represented in the complex plane as shown in §7.4; an example is part A of Figure 7.9, representing such a process at high frequencies. The phase angle is related to the iterated construction of the mathematical set on which the dissipation associated with the process occurs.

Now suppose that a chemical diffusion process develops in the neighbourhood of this interface – this is where we begin to think in terms of hyperfractality. The question arises, what is the relation between the phase angles of the two processes? The approach of §4.2 suggests that they should differ by 45° (this follows from equation 4.60), which indeed is observed in practice.

Figure 7.9 gives an example. We can go further. Controlling the development of the first process so that we can follow that of the second; this is shown in Figure 7.10.

LE MÉHAUTÉ A. and CRÉPY G., *Solid State Ionics 9-10*, 1983, 17.

HAMAIDE T., GUYOT A., LE MÉHAUTÉ A., CRÉPY G. and MARCELLIN R., *J. Electrochem Soc. 136, (11)*, 1989, 3152.

7.11 Use of non-integral derivation in control engineering: the CRONE method

A typical application of non-integral derivation in control engineering is the CRONE – "Commande Robuste d'Ordre Non-Entier" – "Robust Non-Integral Control" – developed by Alain Oustaloup and his group at the University of Bordeaux-I. It has the remarkable property of providing a damping that is insensitive to the parameters of the system being controlled, over a wide range of frequencies: this is the "robustness" of the name.

The operation is described by a differential equation of non-integral order, of the same form as that which describes the ebb and flow of water in a porous barrier. This also is robust in the sense – paradoxical to the mechanical engineer – that the governing law is independent of the mass of water involved in the motion.

The robustness comes from the non-integral derivation operator. Oustaloup has shown that the equation for the water-flow process is

$$\tau^n \mathscr{D}_t^\alpha P(t) + P(t) = 0 \tag{7.39}$$

with $1 < \alpha < 2$. $P(t)$ is the dynamic pressure at the interface with the barrier and $\tau = f(M)$ is a time constant, a function of the mass of water in motion.

If we make a cut in the complex plane along the negative real axis, to take account of the possibility of an infinity of negative real roots, the characteristic equation corresponding to (7.39)

$$(\tau p)^\alpha + 1 = 0 \tag{7.40}$$

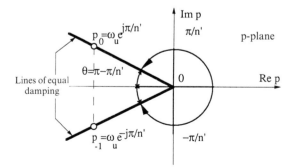

Figure 7.11 Location in the complex plane of the roots of the characteristic equation (7.40) illustrating the robustness of CRONE.

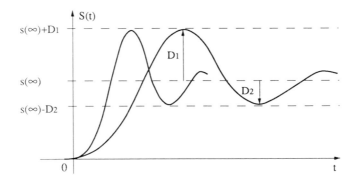

Figure 7.12 Time variation of transients, illustrating the robustness of CRONE in this dimension.

has two conjugate complex roots and the angle between their radii is constant, independent of τ. These are two half lines with equal damping which illustrate the robustness in the plane of p: see Figure 7.11, where $\omega_u = 1/\tau$.

In the space-time plane the robustness is expressed by a response that is equally independent of the parameters of the process. Only the natural frequency varies; the form of the transient is constant, although its timescale varies. This is shown in Figure 7.12.

In p-space the closed-loop transfer function for CRONE is

$$F(p) = 1/([1+(\tau p)^\alpha]$$ (7.41)

The open-loop transfer is

$$F(p) = \beta(p)/[1+\beta(p)]$$ (7.42)

where $\beta(p)$ is the open-loop gain.

In the Black diagram the location of $\beta(p)$ is a line parallel to the imaginary axis with real part, for $1 < \alpha < 2$, between $-\pi/2$ and $-\pi$. Since the closed-loop dynamical behaviour is related to the open-loop behaviour in the neighbourhood of the frequency for which the gain is 1, the robustness of the damping is guaranteed if there is a segment of this line in the neighbourhood of this frequency. Such a segment, shown in Figure 7.13, is called an open loop frequency gauge.

The robustness of the device results from the sliding of this gauge segment along the vertical line when the parameters of the process are changed; the phase

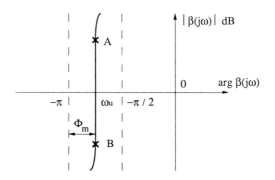

Figure 7.13 Robustness of CRONE in frequency space. AB is the open loop frequency
gauge to be synthesized.

displacement Φ_m (see Figure) remains constant, and consequently the damping
factor also.

Alain Oustaloup has been working on the construction of such "gauges"
since he took out his first patent in 1978; his results show that they require
recursive electronic structures of a fractal nature.

OUSTALOUP A., "Oscillateur sinusoïdal d'ordre demi entier", French patent
78.35728, 1978.
OUSTALOUP A. and NOUILLANT M., *La suspension CRONE*, in press, 1989.
OUSTALOUP A., *IEEE trans. on Circuits and Systems 28(10)*, 1981, 1007.
OUSTALOUP A., *IEEE trans. on Circuits and Systems*, meeting, Chicago, Illinois,
April 1981.
OUSTALOUP A., *Systèmes asservis linéaires d' ordre fractionnaire*, doctoral thesis
Bordeaux, 1981.
OUSTALOUP A., *Systèmes Asservis linéaires d' Ordre Fractionnaire, Théorie et
Pratique*, Masson, Paris, 1983.

Chapter 8

Complexity

There are at least two ways of attacking complexity in physics. One, much the commonest, is to take a process that is already well understood and add further parameters and degrees of freedom, in the hope of matching the behaviour of the process that is not yet understood. The other, much more difficult, is to attempt an analysis from first principles.

In this chapter we aim to show how fractal geometry enables the two approaches to be combined.

8.1 Multi-fractal electrical network

We consider a random resistor network at the percolation threshold through which a current is flowing. If $n(\mu)$ is the number of bonds over which the potential drop is μ the k'th moment of the potential distribution is

$$M_k = \sum_\mu n(\mu)\mu^k \sim L^{-p(k)/v} \tag{8.1}$$

where L is the dimension of the lattice, v is the exponent of the correlation length and $p(k)$ is a function of k.

An optimisation calculation, assuming that the $p(k)$ form an infinite set of independent exponents, leads to the following expression for $n(\mu)$:

$$n(\mu) \sim L^{f(\alpha)} \tag{8.2}$$

where $f(\alpha)$ is the Legendre transform of $p(k)$, $f(\alpha) = k\alpha - p(k)/v$ with $\alpha v = dp/dk$

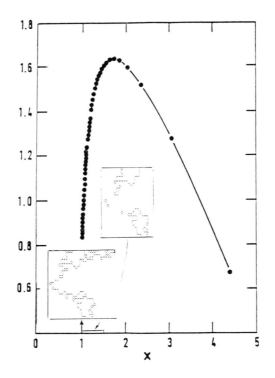

Figure 8.1 Fractal dimension of the potential distribution in a percolation network (130 × 1130) plotted against the ratio log μ/log μ_{max} cf. De Arcangelis.

(cf. Equations 4.29–30). In this form f(a) is the fractal dimension of the set of bonds over which the potential drop is μ, so this dimension is determined by the potential drop. α describes how the potential falls to zero as the size of the network increases.

DE ARCANGELIS L., "Disorder and Mixing", edited by E. Guyon, J.P. Nadal and Y. Pomeau, *NATO ASI*, Series vol. 152, 31.

CONIGLIO A., *Physical 40 A*, 1986, 51.

MANDELBROT B., "Statphys 17" edited by C. Tsallis, in press, *Physica A*.

MANDELBROT B., *Frontiers of Physics, Landau Mémorial Conférence*, edited by E. Gotsman, Pergamon, New York, 1989.

MANDELBROT B., *Pageoph 131(1/2)*, 1989, 5.

STANLEY H.E. and MEAKIN P., *Nature 335*, 1988, 405.

8.2 Fractal molecules and metrical ambiguity

The idea of a fractal molecule was first introduced in the attempt to explain the structure of a polymer chain as a *self-avoiding random walk* (de Gennes 1979). A deeper understanding of fractality led to further ideas; for example, Rouvray and his collaborators have used fractal concepts to describe alkyl molecules. With a rather different approach, Sautet and his collaborators have treated fractal molecules as Cantorian sequences of different blocks and have made some calculations of quantum wave transmittance by such systems – see Figure 8.2.

Energy

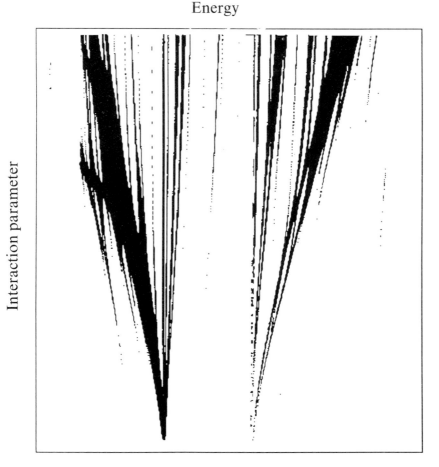

Figure 8.2 Attractor for a quantum wave transmittance through a macromolecule in which a Cantorian fractal section lies between two periodic sections. The pass bands are shown in black.

Work in this field has thrown up a number of mathematical problems that have not yet been solved; we mention a few of these here.

It is known that if a protein chain is stretched into a line and released it folds back into its original 3-dimensional configuration remarkably quickly, in a few milliseconds; whereas an investigation of all the possible ways of folding, to select the one preferred, would take an immense time, possibly thousands of millions of years. So there must be some "engine" that drives the process, not just classical statistics; in a very general way, this suggests the following mathematical problem.

Suppose we have a long cord coloured in sections with n colours; the sequence of colours could be periodic, quasi-periodic or according to some sequence, for example Fibonacci or Cantor; Plate 9 will give an idea of the scale of the problem we are considering. These questions arise:

1. Can the cord be wound up so that each colour constitutes a space of given fractal dimension?

2. Assuming that this can be done, and given the constraints of continuity, can the fractal dimensions be chosen freely or are there relations between them?

3. If there are relations between these dimensions, how are they determined by the distribution of colours along the cord?

4. What relation is there between the decision tree that gives the final structure and the initial colour distribution?

5. What is the minimum number of tests needed in order to determine the n metrics associated with the different colours?

These questions are all expressed in terms of fractality, so it must be possible to formulate a criterion for the organisation in terms of entropy: for example, that the set of fractal dimensions is such that the entropy of the folded cord, in the Mendès-France sense, is minimum.

The next step would be to add enthalpic considerations to those of fractality, with assumptions of interactions between the different colours.

The treatment so far has been purely geometrical, taking no account of the parametrisation of the molecule. This, whilst we may not know just how, must play a basic role in determining the physic-chemical properties of the polymer It is believed, for example, that quantum waves can experience a metrical ambiguity, that is, in a single structure they can have a choice of two values for the fractal dimension and choose a geometry according to one of these and a parametrisation according to the other. It has been shown that certain chemical structures, called *Janals* (see § 4.2.2) can, under certain conditions involving the ratios of the dimensions and the associated "mobilities", display very special

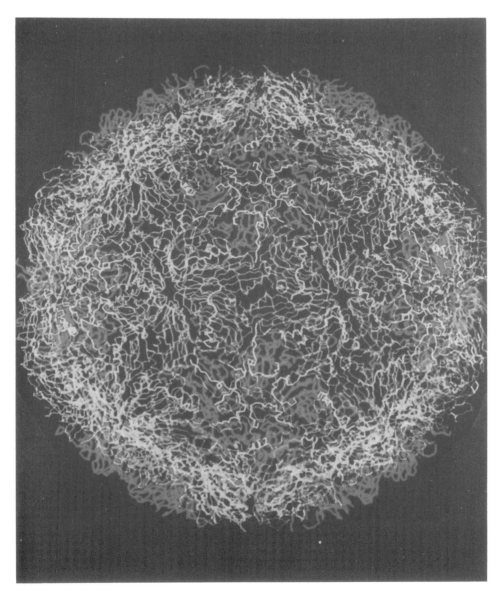

Plate 9. Example of a very complex structure: a virus, the structure determined by synchrotron radiation. (*Synchrotron Radiation News*, p. 24).

properties such as those related to a new form of superconductivity. Further information on these aspects can be found in the references below.

CASSOUX P., KAHN O. and LE MÉHAUTÉ A., French patent 85.10325, 1985.
DE GENNES P.G., *Scaling Concepts in Polymers Physics* Cornell, 1979.
ROUVRAY D.H. and PANDEY R.B., *J. Chem. Phys.* 85(4), 1986, 2286.
LE MÉHAUTÉ A., "Supraconducteur à haute température" E.S. 91653/40041, France, 9 May 1983.
LE MÉHAUTÉ A., "Nouvelle structure de puits quantiques", French patent, 89.16190, 12 December 1989.
LE MÉHAUTÉ A., "From Dissipative to Non Dissipative Processes in Fractal Geometry; The Janals" in press, *New Journal of Chemistry*, 1990.
SAUTET P., HÉLIODORE F., LE MÉHAUTÉ A. and JOACHIM C., in press.

8.3 Julia and Mandelbrot sets, time and stability considerations for iterated functions in the complex plane

Since the work of Poincaré the study of dynamical systems has made use of the results of iterating many times with a functional operator. A particular interest attaches to the neighbourhood of fixed points, that is, of the points for which $f(x) = x$. As well as fixed points there can be cycles, defined by $f(x_i) = x_{i+1}$ for $i = 1, 2, \ldots k, x_{k+1} = x_1$. The *orbit* of x is the sequence of points $x_n = f^{(n)}(x_0)$; this can be a single point, a finite cycle, tend to infinity, be completely chaotic or show any other type of behaviour.

An iteration that has been given much study, in particular by Douadi, by Hubbard and by Mandelbrot, is $f_c(z) = z^2 + c$, where z is a complex variable and c is a complex constant. The first point of interest in connection with this is that for certain values of c the iteration remains finite, that is, the set K_c of points $f_c^{(n)}(z)$ is bounded.

Suppose A_0 is an arbitrary set of points in the plane and A_1 the set that is transformed into A_0 by f_c, that is, the set of points z such that $z^2 + c$ belongs to A_0; we can write this as $A_1 = f_c^{-1}(A_0)$, and in general $A_n = f_c^{-1}(A_{n-1})$. Thus A_n is a decreasing sequence of sets of points, each contained in its predecessor, and the limit of this sequence is the set K_c. The boundary of K_c is called the *Julia set*, J_c.

The sets generated by this process have very different forms according to the value of c. They fall into two classes: they are either connected – "in one piece" – or of the Cantorian "dust" type. As was stated in § 7.6, there is a relation between

Julia sets and the poles of certain electrical circuits that are characterised by scaling laws – this is a subject that merits further study.

The *Mandelbrot set* is the set of values of the constant c for which the set K_c is connected; it can be shown that a sufficient conditions for this to hold is that $|f_c^{(n)}(0)|$ is finite for all n. It is a closed set of particularly complicated form contained within a circle centred on the origin and radius 2 units. It is not self-similar, because the location of any part within the complete set can be deduced from an examination of that part. The local behaviour reproduces in a way the corresponding Julia sets.

BLANCHARD P., *Bull. Am. Math. Soc. 11(1)*, 1984, 85.

BARNSLEY M.F., MORLEY T.D. and VISCAY E.R., *J. Stat. Phys. 40 (1/2)*, 1985, 89.

DOUADI A., "Image des Mathématiques" 25, *Courrier du CNRS*, 1985.

DEVANEY R.L., *Science 235*, 1987, 342.

MANDELBROT B., *The Fractal Geometry of Nature*, Freemann, San Francisco, 1982.

PEITGEN H.O. and RICHTER P.H., *The Beauty of Fractal*, Springer-Verlag, Berlin, 1986.

8.4 Propagation of electromagnetic waves in a multi-layer structure

Suppose we have a multi-layer structure constructed from two generic layers A, B in some way – periodic, quasi-periodic, Fibonacci sequence, fractal etc. The transmission and reflection spectra associated with such a structure will be complex, as in the example of a Fibonacci structure, meaning that successive layers are constructed as A, B, BA, BAB, BABBA, BABBABAB, etc. We put this problem: given the spectrum, and all the necessary electromagnetic properties of the layers, is there any way of inferring the scale laws that govern the multi-layer construction? Apart from the Fibonacci case, we do not know; the integral of the spectrum over the frequency range gives a "Devil's staircase" (cf. §4.1.1, Fig. 4.2), but this identifies neither the order of iteration nor the law.

For this, we have suggested a concept which we call "impedance loss" (IL), with the aim of focussing attention not on the spectrum but on the electromagnetic impedance Z(p), which has the advantage of conserving the information concerning the modulus and phase of the wave over the frequency spectrum. The definition is:

$$V(p) = \int_0^p Z(s)\,ds \bigg/ \int_0^\infty Z(s)\,ds \qquad (8.3)$$

The main features of this function are

– it is physically significant, since it can be associated with the "diffusion" of a pseudo-particle in a heterogeneous medium

– it gives a representation in which each iteration in the multi-layer construction is shown separately

It is thus a true "signature" for the process, since it reveals both the sequence (by its form) and the order of iteration (by its complexity).

Quantum physics

ALI M.K. and GUMBS G., *Phys. Rev. B 38(10)*, 1988, 7091.
ABE S. and HIRAMOTO H., *Phys. Rev. A 36(11)*, 1987, 5349.
GRUMB G. and ALI M.K., *Phys. Rev. lett. 60(11)*, 1988, 1081.
KAHN L., ALMÉIDA N.S. and MILLS D.L., *Phys. Rev. B 37(14)*, 1988, 8072.
KOHMOTO M., KADANOFF L.P. and TANG C., *Phys. Rev. lett. 50(23)*, 1983, 1870.
KOHMOTO M., SUTHERLAND L.B. and IGUSHI K., *Phys. Rev. lett. 58(23)*, 1987, 2436.
KOHMOTO M., SUTHERLAND L.B. and TANG C., *Phys. Rev. B., 35(3)*, 1987, 1020.
KOHMOTO M. and BANAVAR J.R., *Phys. Rev. B 34(2)*, 1986, 563.
LARUELLE F., THIERRY-MIEG V., JONCOUR M.C., and ETIENNE B., *J. Phys Paris Colloque C5.11, 48*, 1987, C5-529.
LARUELLE F. and ETIENNE B., *Phys. Rev. B 37(9)*, 1988, 4816.
MOOKERJEE A. and VIJAI A., *Phys. Rev. B 34(10)*, 1986, 7433.

Electromagnetism

CHABASSIER G., GABRIAGUES J.M., HELIODORE F., LE MÉHAUTÉ A., MOUCHARD J. and SAUTET P., "Dispositif de Filtrage Electromagnétique", French patent, 88.03700.
GUPTA S.D. and RAY D.S., *Phys. Rev. B 38(5)*, 1988, 3628.
HÉLIODORE F. and LE MÉHAUTÉ A., in Abstract of "Progress in Electromagnetic Research Symposium", in press.

8.5 Time and morphogenesis: iterated affine and quadratic transformations

Following the work of Mandelbrot on transformations of the complex plane into itself, a great deal has been done in this field, particularly by Barnsley in connection with image compression. One example he has given is the fern leaf. This can be broken down into four sub-sets; with each is associated an affine transformation, and four probabilities weight the measures of the respective sub-sets; thus the measure of the tail is weighted less heavily than that of the leaves. (See plate 10).

Provided that each transformation is contractive – that is, it always reduces the distance between any pair of points – there can be an attractor for the set; in this case the attractor is simply the leaf itself. Conversely, any decomposition of an image into a finite number of transformations of the type deformation/ reduction leads by iteration to an attractor that is the image itself: this is the principle underlying the compression of a fractal image – only the parameters of the transformation need be transmitted, and not the image itself.

The same method can be applied in attempting to define the invariants for complex phenomena, such as electromagnetic wave propagation in a disorded medium subject to scaling laws. For example, the representation of the relevant IL is an attractor for the propagation of an electromagnetic wave in a medium that is stratified in different ways: could one find a set of transformations that would lead to this attractor? The answer is yes, but the transformations seem not to be affine but possibly quadratic. There is a need for much work here, for it seems that quadratic transforms have a much deeper physical significance than merely reproducing numerical observations.

A further question concerns the stability of such transformations. Experience so far suggests that they are very unstable, and that very small changes to the parameters in a quadratic transformation can result in very large changes to the form of the resulting attractor. This too is a field – fascinating and still largely unexplored – in which much work remains to be done.

BARNSLEY M.F., *Fractals Everywhere*, Academic Press, New York, 1988.
HÉLIODORE F., CHABASSIER G. and LE MÉHAUTÉ A., "Progress in Electromagnetic Research Symposium", Boston, 1989, in press, *J. Electromagnetism Wave and Applications*.

By way of Epilogue

The great value of fractality is that it not only unifies within an elegant theoretical framework a large number of phenomena that cannot be brought within the traditional bounds, but also encourages a return to graphical methods – which in scientific studies had been almost completely replaced by analysis.

Graphics reveal the beauty of fractal geometry, the close relation between

$$\begin{vmatrix} x \\ y \end{vmatrix} \xrightarrow{w1} \begin{vmatrix} x' = 0,5 \\ y' = 0,16y \end{vmatrix} \qquad \begin{vmatrix} x \\ y \end{vmatrix} \xrightarrow{w2} \begin{vmatrix} x' = 0,85x + 0,04y + 0,072 \\ y' = -0,04x + 0,85y + 0,128 \end{vmatrix}$$

$$\begin{vmatrix} x \\ y \end{vmatrix} \xrightarrow{w3} \begin{vmatrix} x' = 0,2x - 0,26 + 0,4 \\ y' = 0,23x + ,22y + 0,011 \end{vmatrix} \qquad \begin{vmatrix} x \\ y \end{vmatrix} \xrightarrow{w4} \begin{vmatrix} x' = 0,15x + 0,28y \\ y' = 0,26x + 0,24y - 0,096 \end{vmatrix} \qquad \text{(a)}$$

(b)

Plate 10. Generation of forms by iteration of affine transformations (8.5). (a) set of four transformations (b) fern leaf generated by iteration of (a).

the multiplicity of scales and the harmoniousness of the figures, as though the mind could understand the generator underlying what is generated, the time underlying the iteration, the unity underlying the multiplicity. Geometry and symmetry have always had an influence on the canons of beauty: historical references to this have been given by Mandelbrot in his article in the *Encyclopedia Universalis* and in his lectures in the seminar on Philosophy and Mathematics at the École Normale Supérieure (*Penser les Mathématiques*, Edition du Seuil 1982).

Certain artists of high calibre are in no doubt of this. One is my friend J.P. Agosti, who wrote in the foreword to one of his exhibitions: "I discovered gardens of dry stones, microcosms of landscapes of seas and islands. This feeling of an internal similarity of space I found at different levels in all the places I visited. I had found the link between the cosmos and its representation in the mind." Fractality haunted Agosti even before he became aware of it and before he knew Mandelbrot.

It is this fundamental awareness that gives Mandelbrot's discovery its interest and importance, beyond its impact on mathematics and physics.

Index